African Agriculture and the World Bank
Development or Impoverishment?

Editorial Group
Kjell Havnevik, Deborah Bryceson, Lars-Erik Birgegård,
Prosper Matondi and Atakilte Beyene

*Report based on a workshop organised by
the Nordic Africa Institute, Uppsala on March 13–14, 2007
with funding from the Swedish International Development Cooperation
Agency, Sida, Stockholm*

NORDISKA AFRIKAINSTITUTET, UPPSALA 2007

NAI Policy Dialogue is a series of short reports on policy relevant issues concerning Africa today. Aimed at professionals working within aid agencies, ministries of foreign affairs, NGOs and media, these reports aim to inform the public debate and to generate input in the sphere of policymaking. The writers are researchers and scholars engaged in African issues from several disciplinary points of departure. Most have an institutional connection to the Nordic Africa Institute or its research networks.

To ensure the actuality and relevance of the topics in these reports, the Nordic Africa Institute welcomes inputs and suggestions from readers in general and policymakers in particular. Please e-mail your comments to: birgitta.hellmark-lindgren@nai.uu.se.

Indexing terms:
Agriculture
Rural development
Sustainable agriculture
Farming
Smallholders
Land tenure
Commodity markets
Poverty alleviation
Structural adjustment
Development strategy
Africa

Language checking: Elaine Almén
Cover photo: Women farmers with their hoes, Kighare, Same, in the north-east of Tanzania, near Kilimanjaro. Sean Sprague, Phoenix bildbyrå.
ISSN 1654-6709
ISBN 978-91-7106-608-4 (print)
ISBN 978-91-7106-609-1 (electronic)
© The authors and Nordiska Afrikainstitutet 2007
Printed in Sweden by Elanders Sverige AB, Stockholm, 2007

Contents

Acknowledgements	5
Introduction	7
World Bank Policy and the WDR 2008	9
African Development Policies over the Last 25 Years	14
Structural adjustment strategy and the weakening of African states	15
Return to a poverty and growth perspective	17
Liberalized market's impact on African smallholder agricultural production	21
Staple food crop performance and rural household food security	21
Global commodity markets	25
Underlying factors	27
The case of cotton	28
Future prospects	29
Land Rights, Markets and Capital	32
Land issues and land tenure reform	32
Land individualisation and markets	34
Large versus small scale farmers – Growing landlessness	35
Foreign land investments	36
Africa's uncapitalized smallholder production: Three decades of declining investment and agricultural extension	37
Institutional Supports for African Smallholder Agriculture	40
African nation-states	40
Their role in agrarian development	40
Disputed character and functioning of African states	41
Resource mobilization capacity of the state	42
Rural households as transforming institutions interacting with market and state trajectories	44

African Rural Agency in Response to Global
Market Pressures 49
    Non-agricultural rural income diversification 51
    Urban migration and employment 52
    Relinquishing smallholder production autonomy:
    Contract farming and agricultural wage labour 53
    Labour uncertainties of smallholder household
    members 54

Reading Between the Lines of the WDR 2008:
African Smallholders' Rural Future 56
    Abandoning food security and smallholder
    agricultural development aims 56
    Sequenced timing of depeasantization 60
    Well-charted predictable developmental path
    or descent into deepening poverty? 61
        African smallholder export and food production 62
        Alternatives to smallholder production 62
        Provision for those without an alternative:
        African 'rural holding grounds' 63

Conclusion 65

Acronyms 68

References 69

Acknowledgements

The Editorial Group would like to thank all those who in various ways have contributed to the realization of this book. We are grateful to the Swedish Agency for International Development, Sida, and the Nordic Africa Institute (NAI) for funding, encouragement and for providing the working facilities. Carin Norberg, the director of NAI, supported the idea and the implementation of the project throughout. A workshop aimed at refocusing attention on African agriculture in the context of the World Development Report 2008 driven by the World Bank was held at NAI on March 14 and 15 2007. The participating researchers from Africa and Europe contributed significantly to the discussion and analysis that formed the framework of the book. This materialized as well because of very competent facilitation of the workshop by Lars-Erik Birgegård. Prosper Matondi prepared a comprehensive report from the workshop and Tania Berger, The Nordic Africa Institute, provided professional support for its planning and implementation.

A number of researchers with long Africa related experience provided specific written contributions to the book including Atakilte Beyene, Stockholm University/Swedish University of Agricultural Sciences, Uppsala, Deborah Fahy Bryceson, African Studies Centre, Oxford University, Peter Gibbon, Danish Institute of International Studies, DIIS, Copenhagen, Kjell Havnevik, The Nordic Africa Institute, Hans Holmén, Linköping University, Prosper Matondi, Center for Rural Development, University of Zimbabwe, Harare, Yenkong Ngangjoh-Hodu, The Nordic Africa Institute, Rune Skarstein, The Norwegian University of Technology and Science, NTNU, Trondheim and Åsa Torkelsson, University of Stockholm. Deborah Bryceson and Kjell Havnevik were responsible for the final editing of the book based on all written contributions and a number of comments.

The Editorial Group is thankful for comments and clarifications to parts of or the full book from Amanda Hammar, Dorte Thorsen, Carin Norberg and Fantu Cheru, all based at the Nordic Africa Institute. In addition Christer Gunnarsson, University of Lund, Peter Hazell, the Imperial College, London, Tore-Linné Eriksen, Oslo University College, Mats Hårsmar, the Swedish Foreign Ministry, Stockholm and Anita Ingevall, the Division of Natural Resources and the Environment, the Swedish Agency for International Development, Sida, Stockholm, all made important contributions.

We are grateful to Elaine Almén for the language editing of the book and to Boel Näslund for preparation of the manuscript for publication. We also wish to thank the publication and information department of NAI for professional support throughout the project.

The Editorial Group hopes that the content and critical reflections contained in this book can stimulate discussions, lead to innovative inquiries and widen the policy space for future pathways for African agriculture and development that can benefit smallholders and rural people.

Uppsala in September 2007

The Editorial Group:
    Kjell Havnevik, The Nordic Africa Institute, Uppsala
    Deborah Fahy Bryceson, African Studies Centre,
        Oxford University
    Lars-Erik Birgegård, Uppsala
    Prosper Matondi, Center for Rural Development,
        University of Zimbabwe, Harare
    Atakilte Beyene, Stockholm University/Swedish University
        of Agricultural Sciences, Uppsala

Introduction

Agriculture's dominant role in Sub-Saharan Africa's local, national and regional economies and cultures throughout pre-colonial history has been foundational to 20th century colonial and post-colonial development. No other continent has been so closely identified with smallholder peasant farming. Nonetheless, smallholder farming has been eroding over the last three decades, perpetuating rural poverty and marginalizing remote rural areas. Donors' search for rural 'success stories' merely reinforces this fact. The current role of agriculture and rural development in African national economies and its potential for improving material standards of living and life chances is thus of pressing concern. It is time to ask if agriculture spells welfare enhancement or decline for Africa's rural dwellers. Certainly many farmers have voted with their feet by increasingly engaging in non-agricultural livelihoods or migrating to urban areas. In so doing, the significance of agriculture for the majority of Africa's population has altered.

The World Bank has played a prominent role in shaping agricultural policy in Africa for the last thirty years. Its insistence on structural adjustment programmes in the aftermath of the oil crises of the 1970s reversed previous development investment in peasant agriculture. Today's African agricultural sectors demonstrate the ambiguous outcomes of the policy trajectory set on course by the World Bank in the early 1980s. It is for this reason that the World Development Report (WDR) 2008 thematically focussed on agriculture is of special interest. It examines agricultural development worldwide, comparing African agriculture relative to performance elsewhere.

This paper offers a critical reflection of the WDR's portrayal of world agriculture with respect to Africa. We present an overview of African land, labour and capital market dynamics since the oil crises of the 1970s, contextualising the current institutional state of play. Examining three decades of agricultural decline in Sub-Saharan Africa, we also highlight the role of the

World Bank in determining the relative roles of the state and private sector and agricultural output trends. Farmers' economic and social choices are highlighted before probing the central issue facing Africa's rural dwellers, namely the increasing displacement of their agrarian labour. We ask what the implications are of the World Development Report 2008's recommendations for the survival of smallholder farmers. In the concluding section, measures to raise agricultural productivity and reduce rural poverty are suggested to invigorate, rather than marginalize, African family farming.

World Bank Policy and the WDR 2008

Over the past 50 years of its existence, the World Bank has become the central international agency prescribing economic development policy to the world's nation-states. It was strategically placed to offer advice to Sub-Saharan African countries after their achieving of national independence in the 1960s. Under structural adjustment conditionality of the 1980s continuing to the present, the World Bank's prescriptions have become largely mandatory for the debt-ridden national economies of the continent. Its influence over a country's policies is generally in direct inverse proportion to that country's economic strength. Thus, most African countries have to greater or lesser degrees espoused and implemented World Bank development policy for the last 25 years, and African agricultural sectors, in effect, demonstrate through continuous low growth rates and deepening rural poverty, the impact of World Bank policies.

A recent evaluation[1] of the World Bank's research output challenged the institution's reputation as the world's 'knowledge bank' referring to its habit of taking 'new and untested results as hard evidence that its preferred policies work', singling out the flagship World Development Reports published annually as a medium through which advocacy of the World Bank's favoured policy recommendations sometimes takes precedence over balanced analysis.[2] So where does the World Development Report 2008 fit

---

1. 'An Evaluation of World Bank Research 1998–2005' (September 2006) chaired by Angus Deaton (Princeton University) and a team of other academics. This report is highly significant for being the first of its kind in over 20 years. See The Economist, 'What the World Bank knows', 13 January, 2007.

2. A number of critiques of the World Bank's power to influence policy on untested ideological grounds have surfaced in recent years (Fox 2002, Peet 2007, Broad (2006) in addition to the insights provided by Stiglitz (2002), a former Chief Economist at the World Bank. These have been long preceded by trenchant criticism of World Bank

into this schema? On the face of it, the WDR 2008 espouses a continuation of World Bank rural policies of the last quarter century. First, it argues that agriculture is key to poverty alleviation, especially for African smallholder farmers. The majority of Africa's poor live in rural areas and farm to varying extents. Agricultural growth has unique potential to alleviate poverty in agrarian-based economies.

Second, it stresses that liberalized national markets will remain the primary force for achieving productivity increases and poverty alleviation. Accelerated growth will be achieved through agricultural productivity improvement but the 'green revolution' model of state investments and subsidized support for agricultural inputs are discounted. African states are seen to be seriously flawed and therefore best restricted in scope and decentralised to preclude government intervention in the national economy. Smallholder households will participate in commodity, capital, land and labour markets, to seek multiple pathways out of poverty; either through encompassing agricultural production, rural non-agricultural enterprises or out-migration.

Beneath these entirely business-as-usual policies, there are starkly contradictory objectives: the humanitarian concerns of poverty alleviation clash with a Darwinian market fundamentalism.[3] Will African peasant farmers' lot improve or decline further? The report has a casual way of not distinguishing the radically different policy needs of small as opposed to large-scale agriculture. In global agricultural commodity markets, African smallholder producers have been losing market share continuously over the last three decades. Africa's traditional export crops, the beverage crops: coffee, cocoa, tea, as well as cotton, tobacco, cashew, etc. have steadily declined to now quite negligible export levels. The comparative advantage that African smallholders held in these crops has been undermined by far more efficient producers elsewhere. There is no evidence provided to suggest that the broad masses of African small-scale peasant farmers will experience anything other

---

policies by the Organization of Africa Unity (United Nations Economic Commission for Africa 1989) and UNCTAD.

3. Market fundamentalism is here defined as the unshakeable belief in the innate nature of the market as a prime mover of exchange and optimizer of production without regard for the political imbalances and social biases of markets as historical institutions. States are seen as potential concentrations of vested interests and power in stark contrast to markets as neutral forums of exchange.

than continuing difficulties in meeting the rigours of global commodity market chains with their highly regulated standards and time schedules.

Large-scale producers, on the other hand, composed of small African rural elites of capitalized farmers and relatively limited numbers of foreign-owned plantation and estate owners are likely to not only cope but also flourish. Reading between the lines, and in places boldly stated, is the World Bank's endorsement that productive agriculture in the 21st century is inevitably large-scale and will prevail over uncompetitive small-scale producers.

Paradoxically, the World Bank has a long tradition of championing smallholder farmers. Structural adjustment policies were implemented in the name of 'getting the prices right' to promote market efficient resource allocation for the benefit of smallholders. Consistently World Bank agricultural policies have displayed contradictory tendencies and a glaring discrepancy between stated objectives and actual outcomes. Nonetheless, the World Bank has rarely been held to account. Peasant farmers have been too dispersed and without a voice whereas heavily indebted African governments are too dependent on the World Bank's conditional aid to criticize the policies it enforces.[4]

African agriculture was in the World Bank's spotlight 25 years ago with the publication of the Berg report entitled *Accelerated Development in Sub-Saharan Africa: An Agenda for Action* (1981) and the *World Development Report 1982* on the theme of agriculture. These reports identified African state policy intervention, particularly in the form of producer subsidies and parastatal marketing, as key problems to resolve in order to achieve higher agricultural productivity. The encouraging improvements in maize yields from the improved input and fertilizer packages that several African governments were distributing on a subsidized basis went unacknowledged, while the dramatic change in terms of trade following the oil crises of 1973/74 and 1979 and the subsequent world market economic shocks that the continent experienced were largely sidestepped – internal rather than external causes of the African economic crisis were stressed.

In the aftermath, as African countries one by one fell into heavy debt and SAP conditionality was imposed, the budding of a potential green revolution blooming fostered by policies of several African states during the 1970s

---

4. Factions of African governments and elites have also been benefiting from such policies and hence shown willingness to accept them.

was nipped in the bud (Eicher, 1995 and 2001, Eicher and Kupfuma, 1997). Unlike the green revolutions of India, Indonesia and the Philippines, which had afforded their farmers several years of state-supported input subsidy, Africa's green revolution was stillborn (Djurfeldt et al., 2005).

There is one notable concession in the WDR 2008. African smallholders may be allowed 'smart' producer subsidies, which must be restrictively targeted and delimited primarily to fertilizer. Considering that farmers in OECD countries have kept their agricultural subsidies relatively intact throughout the last 20 years as African farmers saw their far more modest subsidies whittled away, this is a small consolation. The average support to OECD agricultural producers fell from 37 per cent of gross value of farm receipts in 1986–88 to 30 per cent in 2003–2005. While this represented a 7 per cent decline, the total amount of support increased over the same period from $242 billion a year to $273 billion a year (WDR 2008, ch. 4, p. 134, July version).

Reviving African attempts to rekindle African green revolution efforts, is ruled out. The World Bank's refusal to endorse a concentrated state-coordinated and international donor supported effort to raise African productivity is likely to preclude the African rural poor's agriculture from expanding beyond basic subsistence. Rather than poverty alleviation, the likely outcome is further impoverishment and rising demoralisation on the part of African farmers who have faced deteriorating production and market conditions and struggled largely unaided for the last 25 years. Under these circumstances there has been growing interest among smallholders in low-input and alternative agricultural methods and technologies including composting and organic cultivation. This is a positive development in view of the long run unsustainability of using fossil fuel based agricultural inputs.

Economic growth, the propulsive force for improving material standards of living, is not on the drawing board for African agriculture. Rather the goal is poverty alleviation. Economic growth is projected to happen elsewhere – in other sectors on other continents, not in African peasant agriculture. Despite the title of the WDR 2008 – 'Agriculture for Development', this document spells the 'end of development' for African smallholder farmers. World Bank technocratic spin is deployed to provide optimism to describe what is an increasingly constricting economic reality.

Under current market fundamentalist thinking, large-scale agriculture is deemed to be competitive, not small-scale family production. African smallholders, therefore, have a 'loser' status, but the World Bank appreciates that allowing the global market to fully decimate African peasant agriculture would spell political and human disaster in the weak African national economies where farmers' only option is to join over-crowded rural and urban informal sectors where average levels of capitalization, skills and productivity are exceptionally low. Thus the African countryside of the future is in effect relegated to a large 'holding ground' to ensure the basic welfare of the rural population and provide labour for other sectors of the economy as and when needed.

The following sections describe the unfolding policies and analyse the major trends that have moulded African peasant agriculture to date.

African Development Policies over the Last 25 Years

Most African countries achieved independence in the 1960s amidst a world commodity boom that encouraged an optimistic belief that Africa could follow Europe with a 'big push' towards modernization and industrialization. Two major contending theoretical and development policy directions emerged during the 1970s. The first argued the need for income redistribution, employment generation, education, health, poverty reduction and environmental and basic needs investment. This approach was reflected in the World Bank's poverty strategy launched in 1973 and the ILO strategy for employment, growth and basic needs of 1976. The vehicle designated to promote this poverty and basic needs orientation was integrated rural development programmes, where agricultural modernization was combined with the supply of physical and social infrastructure. The state had a key role to play in supporting smallholder farmers through agricultural marketing boards and crop authorities. Pan-territorial price systems were applied together with purchase guarantees to provide price stability and reduce uncertainties for smallholder farming. Theoretically these policies were linked to a theory of 'redistribution with growth' (Chenery et al., 1974). The second approach applied neo-classical economic analysis to development problems, underlining the efficacy of product and factor markets for the allocation of resources. This paradigm vied against the basic needs approach for influence after the first oil shock of 1973. The latter gained legitimacy with the emergence of the UN negotiations for a New International Economic Order (NIEO) to balance political power and economic justice between the North and South.

Smallholder peasant agriculture, foundational in most African countries, was seriously challenged by the oil crisis of the 1970s. The problems of rising transport costs, which constituted a growing percentage of free on board (FOB) prices, undermined the competitiveness of smallholder production.

A chain reaction ensued: crop-parastatals' inability to cope with the rising costs of crop transport resulted in late or missed payments for farmers' crops. Faced with declining terms of trade, smallholders' production of export crops started contracting. This engendered an increasing balance of payments problem for the poor non-oil producing countries of the African continent and over-valued exchange rates. Government services and infrastructural provisioning to rural areas slid and a vicious downward spiral set in. The expansive development plans of African states could no longer rely on the expropriation of an agricultural surplus, and became instead increasingly entwined with and dependent upon development assistance and loans.

The election of Reagan in the US and Thatcher in the UK, finally tilted the development paradigm in favour of the neo-liberal approach. In the early 1980s, the US assisted by Great Britain, Germany and Canada initiated a drive towards 'aid coordination', a concept which first surfaced in the OECD in 1981. Two years later, the DAC member countries codified their adherence to a system where the World Bank, the United Nations Development Programme and other lead agencies, such as the IMF; would direct the donor community in order to achieve 'consistency between donor aid policies and programmes and the recipient nations' over-all and sectoral development objectives and needs (DAC, 1983). This implied that developing countries came to form a unified front directed by the IMF and World Bank. But it also led to a major enhancement in the influence of the World Bank and a corresponding increase in the interest of the OECD countries in the World Bank's development policies (Gibbon et al., 1993).

Structural adjustment strategy and the weakening of African states

The World Bank's Berg report of 1981 provided the analytical perspective for implementation of structural adjustment policies (SAPs) in Africa. Stagnant and deteriorating economic conditions in the continent were mainly seen as a product of distortions in local economies brought about by inappropriate government policy interventions. The World Bank and other donors showed no capability of developing a balanced or honest criticism of this model or their own role in promoting it. The IMF, the World Bank and major donors embraced the prevailing neo-liberal climate thereby denying any significant future role for the state in the development process. Instead, the state was seen to facilitate the release of market forces and the 'unchaining' of the private entrepreneur. This was a new term for the African smallholder, who in the modernization paradigm had been viewed as backward and traditional. By designating the rural social context as 'private', the agricultural and rural strategies failed to understand that rural societies were embedded in complex indigenous systems of reciprocity, redistribution and market exchange which encompassed other institutions besides the market (Berry, 1993 and Havnevik et al., 2006). The interactions of these exchange systems and the challenges they posed for an expansion of market exchange based on formal institutions was largely ignored.

Hence, the structural adjustment programmes aimed at establishing and supporting formal institutions and included producer price reform, removal of subsidies, liberalization of internal and external trade, new foreign exchange regimes premised on severe devaluations, cost-sharing for state-supplied services, privatization and contraction and restructuring of government institutions. A number of these reforms would, in any case, have had to be implemented by African governments in the wake of the oil shock, faulty development assistance and a weakening state. However, other elements of the reforms simply reflected the growing sway of market fundamentalism in the most powerful developed countries.

Aid coordination and structural adjustment in Africa were accompanied by 'conditionality', i.e. concessional finance to compliant adjusting countries. A survey covering 1985–87 showed that 'strong adjusters' received an annual increase in concessional finance of 19 per cent, while 'weak adjusters', suffered a decrease of 4 per cent per annum over the same period (UNDP/World Bank, 1989).

By the end of the 1980s it was readily apparent that the structural adjustment strategy was not delivering on its promises. In 1989 a modification of the World Bank's strategy for agricultural development emerged with the report, *Sub-Saharan Africa: From Crisis to Sustainable Growth* (World Bank, 1989). This document acknowledged that prices were important, but only as part of a generally 'enabling' environment of which the key feature was the mobilization of the private sector and the role of the state as an efficient infrastructural provider. This included the provision of market and price information, promoting private and cooperative marketing, building market infrastructure, promoting quality control and establishing a legal framework, including reforms of land laws to secure better individual rights etc. The emphasis on research and extension was carried forward from the 1981 report, but its weight in the 1989 report indicated that these were areas that had experienced severe cutbacks and neglect.

Return to a poverty and growth perspective

The 1980s witnessed stagnant or declining agricultural productivity and deepening rural poverty. The removal of smallholder farmers' input subsidies was a source of widespread resentment for smallholder farmers who watched their yield gains from fertilizer and improved seed usage disappear.[1] Bilateral donors, UN agencies and NGOs began pressurizing the Bank to revise its agricultural policy to more seriously address rural poverty in Africa (UNICEF, 1987, Havnevik, 1987). The World Bank initially responded by adding a social action programme to its second generation of structural adjustment programs, but without altering the central mechanisms of the SAP model. Second, a 'Social Dimensions of Adjustment' project was initiated by the Bank accompanied by statistical exercises, revealing a limited understanding of the African poverty context (Gibbon, 1992). Thereafter, the Bank conducted poverty assessments in a number of African countries and had to concede that it had limited knowledge of the complex nature of poverty (Havnevik, 2000). To address this situation it commissioned a global study of poverty using qualitative as opposed to their usual preference for quantitative data-gathering techniques. The study encompassed interviews with 20,000 poor people, including youth and children, across 23 countries. In the publication, *Voices of the Poor – Crying out for Change* (World Bank 2000a), ten interlocking dimensions of powerlessness and ill-being were identified that the Bank stressed were 'based on the experiences, aspirations and priorities of the poor people themselves' (ibid. p.3).

The first draft of the *WDR 2001* on poverty highlighted smallholders' and poor people's empowerment and security over a more growth-oriented perspective. This position proved highly controversial in the World Bank and the IMF. The draft WDR 2001 argued that effective safety nets should

---

1. This general development could be contrasted with that of Zimbabwe in the early 1980s where state support for smallholders was manifested in growth.

be in place before free market reforms were pushed to preclude the 'losers' from market reforms having no support to fall back on. However, this line of argument was omitted from the final published version of the report.[2]

It was at this juncture that the World Bank's contradictory two-pronged strategy of expressing humanitarian concern for the poor and simultaneously pursuing 'sink-or-swim' market liberalization policies congealed. The published form of the *WDR 2001* and the deluge of descriptive case study data from the *Voices of the Poor* lacked a sense of interconnection and critical assessment of overall social patterns and medium to long-term economic trends. The Bank had repackaged neo-liberalism in a new post-modern liberal form with a nuanced commitment to poverty-alleviating welfare measures while backpedalling on African prospects in world markets (Bryceson and Bank, 2001). An acknowledgement of the inevitability of a marginalized poor under global market liberalization was masked by a populist call for the impoverished to seize 'market opportunities'. In other words, World Bank policy spin stressed the market lottery's lucky handful of winners in rural Africa, rather than drawing attention to the plight of the massive number of losers.

The WDR 1995 entitled *Workers in an Integrating World* had diagrammatically made the situation far clearer. In the log-scale histogram (Figure 1, see below), African smallholder farmers were at the bottom of the world employment pyramid with a 60:1 ratio between OECD earnings at the top and African smallholder incomes at the bottom. Projections for 2010 showed this gap widening in a divergent scenario to 70:1 or narrowing in the convergent scenario to 50:1 (Figure 1).[3] In the WDR 2001, the existence of the rural poor was also dramatized, but emphasis was placed on their 'crying voices' not their muscle power to change the situation. 'Helping the poor' became the foundation for the World Bank's new form of aid conditionality.

---

2. The US Treasury did not accept this approach and key members of the WDR 2001 production team left the World Bank in protest (Wade, 2001). The published version of the WDR 2001 accommodated most of the views of the US Treasury.

3. Indicators so far suggest that the gap has been widening. World Bank data shows that the ratio between OECD and Sub-Saharan African GDP per capita was 46:1 in 1992 and 51:1 in 2005. If the earnings gap between OECD workers and African smallholders increased in direct proportion to these figures it would be roughly 66:1 in 2005 with little stopping it from the predicted 70:1 ratio of the divergent scenario in 2010.

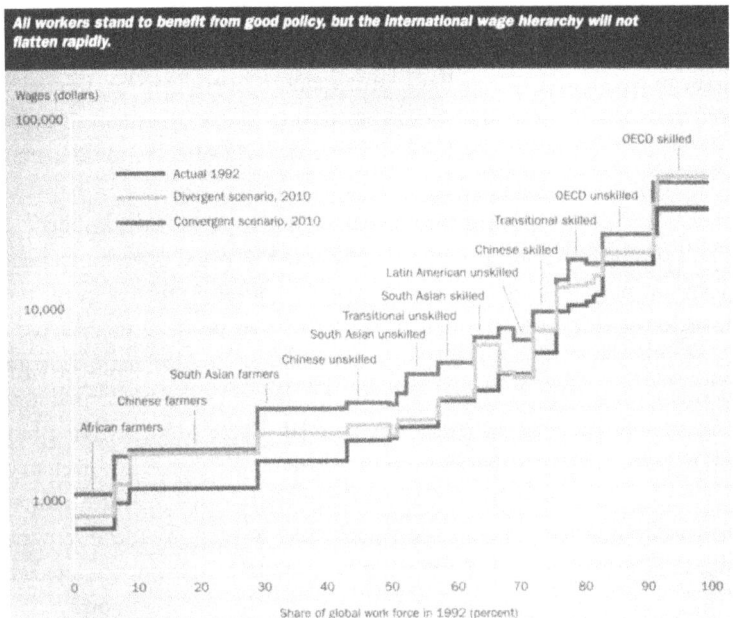

Figure 1: Global labour force and comparative wage levels.
Source: World Bank 1995, p. 121.

A framework for debt cancellation of highly indebted poor countries (HIPC) was linked to a process of national participatory investigation of poverty and the national formulation of poverty reduction strategies. Remarkably, although the focus was the poor and rural areas were identified as where the poorest of the poor were concentrated, there was little attention to agricultural investment. The Millennium Development Goals targeted health and education sectoral investment. In place of smallholder agricultural development policies a diffuse array of rural development policies were advocated. Efforts to strengthen the state through 'election democracy' were combined with decentralization of state functions and a populist narrative of participation and empowerment. A rural community approach became popular among donors and governments. World Bank loans for rural development were channelled to community based projects and activities with

19

a welter of different agendas. In addition the Paris Declaration of 2005 on aid effectiveness exerted pressure for recipient ownership and quality enhancement of aid. World Bank post-modern liberalism meandered with vague calls for poverty alleviation but devoid of clear economic development policies, while continuing to pursue the neo-liberal momentum that was by then well entrenched in African economies.

The WDR 2008 – Agriculture for Development stresses the importance of agriculture for pro-poor growth, accompanied by good governance, decentralization, participation and organizational empowerment of rural people alongside the key role accorded to the private sector, a jumble of old and new themes none of which adds up to a coherent development strategy for smallholder farmers. They are, after all, the losers in the world commodity market competition of the past three decades. There is more than a hint in the report that many of them should leave farming, migrate and seek market opportunities elsewhere because those that remain will constitute a humanitarian welfare cost to their nation-states and possibly to the international donor community. In other words, African smallholder farmers are falling off the graph (Figure 1). They are redundant to the world economy.

The above review of policies has outlined how African countries have followed orthodox World Bank policy strategies over the last 25 years since the 1979 oil crisis and their impact on rural smallholders' welfare. By contrast, countries that have pursued strategies blending protectionism, state subsidies to health and education and exports (e.g. Tunisia and Vietnam) have been remarkably successful in promoting rural economic growth and poverty. In fact, these findings emerge from the World Bank study of reforms during the 1990s (Zagah, 2005). In this document it is acknowledged that the Bank placed too much emphasis on static efficiency, overcoming imbalances and filling gaps with too much eagerness for pushing back the state, and too little weight accorded to achieving long-term productive potential. Nonetheless few of these admissions are embedded in the policy recommendations of the WDR 2008.

How would the African countryside look today if the embryonic green revolution development that was initiated in the 1970s had not been short-circuited by structural adjustment and economic liberalization policies? It is useful to analyze in more detail what happened to African smallholder agriculture after that critical juncture. The next section spotlights the experience of Tanzania.

Liberalized market's impact on African smallholder agricultural production

This section's aim is to consider in more detail how post-SAP agriculture, largely unprotected and unsubsidized relative to agriculture in most other parts of the world, fares in the context of economic liberalisation and the global market. Empirical case study evidence of the impact of market liberalization on smallholders' staple food crop output, notably maize, in Tanzania is examined followed by a broad analysis of African smallholders' prospects in global agricultural export markets.

Staple food crop performance and rural household food security

The implementation of economic liberalization policies during the 1990s was accompanied by the argument that deregulation of prices and free market competition results in the 'right' input prices, and higher producer prices for farmers, spurring them to increase efficiency, produce more, and make investments to raise land and labour productivity. To assess how true this is, Table 1 shows the ratio of farmer producer prices relative to fertilizer input prices for Tanzanian smallholder farmers' four main food crops revealing a decline of the price ratio to the disadvantage of farmers by between 74 per cent for maize and 47 per cent for wheat from 1985–89 to 1998.

TABLE 1: Ratios of average crop producer prices to farmgate fertiliser prices, Tanzania 1985–1998

	1985–89	1990–94	1995–98	1998	% change between 1985–89 and 1998
Maize	1.40	0.83	0.37	0.36	– 74.3
Paddy	2.23	1.39	0.56	0.60	– 73.1
Wheat	1.58	1.87	0.92	0.84	– 46.8
Millet/sorghum	1.05	1.15	0.85	0.54	– 48.6

Source: World Bank (2000b: 46)

The removal of peasant farmers' fertilizer subsidy caused fertilizer prices to rise sharply relative to maize producer prices leading to an 80 per cent reduction of the real return per 'person-day' of maize production from 2,496

TShs at 1998/99-prices in 1992 to 501 TShs in 1998 (Delgado et al., 1999: 95).[4] Under these circumstances it was no longer economically feasible for farmers to use fertilizer (Hawassi et al., 1999). A relative stagnation of the producer prices for maize (and other food crops) contributed to this price squeeze on smallholders.

Table 2 shows the development of real producer prices from 1981 to 1999. For all crops reported in Table 2, there was a rise of the producer price in the early 1990s, which peaked around 1993–94. But between the mid- and the end of the1990s, the real producer prices of all crops have declined. It is noteworthy that the country's most important basic staple food crops, maize and rice, have experienced the largest decline in real producer prices.

TABLE 2: Real producer prices for main food crops, Tanzania 1981–1999 (TShs per kg at 1998/99-prices)[1]

Year	Maize	Paddy	Wheat	Millet	Beans
1981–1985[2]	140	232	195	117	334
1986–1990[2]	149	250	170	109	369
1990/91	106	212	473	279	471
1991/92	279	370	495	289	508
1992/93	298	491	525	365	533
1993/94	256	424	497	376	712
1994/95	181	254	452	484	797
1995/96	165	216	423	538	571
1996/97	138	245	362	245	475
1997/98	117	195	272	175	431
1998/99	118	151	228	175	317

[1] Nominal prices deflated to constant 1998/99-prices using the National Consumer Price Index.
[2] Official procurement prices (before deregulation of prices in July 1990). Reference is to fiscal years (1 July–30 June) coinciding with crop years.
Source: World Bank (2000b:26).

---

4. Assumes hand-hoe technology, involving 123 man-days of family labour, with an average yield of 1,500 kg/ha.

Deregulated staple food markets can be problematic for farmers and consumers alike. Farmers can find the short- to medium-term fluctuations of producer prices in a deregulated market as trying as a long-term downward price trend. Market demand for staple grains in countries where cheap staples comprise the bulk of people's diet is relatively constant whereas supply will fluctuate from year to year based on climate and a variety of other factors. Even rather modest changes in supply can lead to quite large producer and consumer price changes. Before the deregulation of prices in 1990, such variations were modified through the government's price setting. The agricultural producers were informed at planting time on the procurement prices for the next harvest. There could be considerable price changes from one year to the next, but at planting time there was no uncertainty among smallholders about the producer prices of the next harvest (Skarstein, 2005).

Before liberalisation, real producer prices and levels of maize production were positively correlated. They then switched to a negative correlation in the liberalisation period (Bilame, 1996). After liberalisation, high prices reflect a situation of post-harvest deficient supply, while low prices accompany a bumper harvest. Such price variations affect the production plans of surplus producing smallholders. When prices are low in one harvesting season, smallholders tend to make plans for lower marketed output of the crop in question in the next season, and vice versa. This behaviour reinforces the volatility of prices from year to year. In the absence of price stabilisation measures, strong price volatility and stagnation of marketed output is likely to become a basic feature of the Tanzanian maize market, as well as of other deregulated African markets for food grains. Furthermore, both producer and consumer prices exhibit considerable seasonal variability, being lowest just after the main harvest and highest before the next main harvest (Table 3).[5] In the years 1994–1998, the highest monthly producer price of maize was on average 1.8 times higher than the lowest producer price in the same year. The seasonal pattern of consumer prices is much the same, with the notable exception of the year 1994.[6]

---

5. In Zambia and Malawi, the producer prices before a new harvest are generally about two or more times higher than towards the end of the preceding harvest (cf. Øygard et al., 2003).

6. The deviation in 1994 was mainly due to large imports.

TABLE 3: Highest and lowest monthly producer and consumer prices of maize

Year	Producer prices (TShs /kg)			Consumer prices (TShs /tin*)		
	Highest price (H)	Lowest price (L)	H:L	Highest price (H)	Lowest price (L)	H:L
1992	58.30	44.91	1.30	1195.1	945.2	1.26
1993	66.01	39.98	1.65	1354.1	794.3	1.70
1994	128.95	49.90	2.58	1458.2	1104.5	1.32
1995	71.50	49.12	1.46	1695.2	1144.5	1.48
1996	101.76	54.11	1.88	2159.0	1259.5	1.71
1997	120.77	84.96	1.42	2531.3	1795.1	1.41
1998	116,70	61.04	1.91	2924.4	1471.4	1.99
Average 94–98	107.94	59.83	1.80	2153.6	1355.0	1.59

* One tin is approximately 20 kg.
Source: URT/MAC (2000:39) and Marketing Development Bureau (MDB) statistics compiled by D. Rweyemamu of the Economic and Social Research Foundation, (ESRF), Dar es Salaam.

Seasonal price variability is caused by the low demand elasticity of maize. However, in a deregulated market, price variability is reinforced by speculative behaviour among traders. A rising price, which may be triggered by a bad harvest, can result in increased revenue to traders who withhold grain from the market, i.e. postpone sales, when the price is rising. Similarly, wealthy consumers may hoard staple grain in such a situation, if they have the facilities to do so. This behaviour reinforces the seasonal food price rise. Conversely, a declining price, which may be caused by a bumper harvest, will make traders reluctant to buy crops from the smallholders in expectation of an even lower producer price, while selling their stocks in order to avoid future losses. The net effect of widening seasonal food price fluctuations is exposure to food insecurity for poor consumers and lower farmer producer prices thereby decreasing farmers' incentive to produce food crops for commercial markets. Given staple foods' status as an essential basic need, both of these tendencies are likely to have short, medium and long-term detrimental impacts on poverty levels in Tanzania.

For lack of money as well as storage facilities, many poor smallholders engage in 'forced commerce', selling so much of their crop at low prices at harvest time that they do not have enough food grain to cover the needs of their households until the next harvest. They sink further into a poverty trap when they must later on in the season buy food grain at high prices – often with expensive credit – in order to survive.[7] This 'forced commerce' implies a serious income loss to smallholders and a corresponding income gain to private traders (Bhaduri, 1986). Government control of producer prices and public buffer stocks (buying above market prices at harvest time, selling below market prices in the months before next harvest) have so far proved to be the most effective means to alleviate this problem in Africa and Asia (e.g. Gabre-Madhin et al., 2003). The World Bank's insistence on market liberalization precludes such poverty prevention policies. Large numbers of rural people can fall into the vicious poverty trap of basic food shortfalls due to the oversale of their grain stocks and indebtedness to secure purchased supplies at inflated prices.

Having examined the market hazards Tanzanian smallholder farmers face in national grain markets it is now useful to turn to the nature of African liberalized commodity, land and capital markets that confront smallholders.

Global commodity markets

A majority of African countries are commodity-dependent, in the sense that 50 per cent or more of their exports are composed of non-oil commodities. This situation has not changed to a significant extent over the last thirty years. Therefore, developments in international commodity markets are of fundamental importance to Africa's economic prospects.

The period since 2002 has seen the first so-called 'commodity boom' on an international scale since the 1970s. The indexed prices of minerals, ores and metals rose by 100% during the period 2002–05, while the indexed price of crude petroleum rose by 114 per cent (UNCTAD 2006:17). In the case of many minerals, the trend has continued upwards in 2006 and 2007. However, most commodity-dependent African countries rely on agricultural commodity exports and the price trend here has been quite different.

---

7. This is not at all a question of market efficiency, but a question of how a market necessarily works within a particular structure of production.

Despite strong increases from around 2002 in the nominal export prices for a wide variety of agro-commodities, the clear overall trend between 1993–95 and 2003–05 was one of a fall in real prices. Gibbon (2006), analyzing COMTRADE data for 17 groups of agro-commodities over this period, reports nominal price increases exceeding 5 per cent in the case of only two – fresh and chilled vegetables and soybeans. There were nominal price increases of less than 5 per cent for four other commodity groups (beef, tea, cotton and bananas) but nominal price declines in all of the remainder. Decline in nominal prices applied not only to commodities such as coffee where global demand was stagnant, but even for ones such as rice, chickens and cut flowers, where in each case internationally traded volumes increased by more than 40 per cent between 1993–95 and 2003–05.

The main reason for low agro-commodity prices is structural over-supply, especially of undifferentiated basic products. Over-supply, or at least over-capacity, applies in respect of almost all the commodities discussed here, although it is more acute for traditional 'tropical' products. The origins of this are manifold. On the supply side the collapse between 1989 and 1999 of International Commodity Agreements regulating price and volume of products circulating on the world market led to increases in production by the leading existing players (Gibbon and Ponte, 2005). Secondly, in the cases of meats, grains, sugar, oilseeds and cotton, producing countries in the developed world have stimulated over-supply as a result of domestic subsidy systems. Thirdly, over-supply reflects developments on the demand side. In some cases declines in consumer demand have occurred due to health concerns (sugar and beef in developed countries). In other cases it results from technological changes allowing increased substitution (tropical timber, cocoa) or reductions in raw material requirements (tea), or increased ability to use raw materials of lower quality (coffee, tea, cocoa) (Peck, 2001, Oxfam, 2002, Kox, 2000, van Dijk et al., 1998). Finally, there have been large productivity gains for crops such as corn, rice, sugar, soybeans and coffee following propagation of new higher-yielding crop varieties and greater farm mechanization (e.g. Fernando-Cornejo and Caswell, 2006, Gudoshnikov et al., 2004, Childs, 2005). These productivity gains are in turn associated with spectacular increases in production by a handful of countries such as Brazil and Vietnam.

As a result, a polarization is occurring in developing country participation in agro-commodity markets. At one pole are a few countries prominent in export both of those bulk products for which demand is increasing (such as animal feeds) and of higher value commodities including horticulture and aquaculture products. A second group of countries are mostly specialists in traditional tropical products, for which demand has been flat in recent years. This group of countries, which includes the 35 or so commodity-dependent countries in Sub-Saharan Africa, have lost their traditional cost and quality advantages even for traditional tropical products, while figuring very little in the trade for those products that are in high demand. Thus, their share of world commodity trade is declining, alongside their capacity to diversify into higher value commodities or manufacturing.

• *Underlying factors*
The underlying factors driving this polarization are new economies of scale relating to segmentation of demand and the restructuring of global value chains. Mainly in respect of higher-value agro-commodities but also in relation to traditional tropical crops such as coffee, end-markets in developed countries have witnessed a consumer-led proliferation of product and process standards. Because the costs of conformity to these standards are typically physically indivisible, operators benefiting from greater economies of scale in production and post-harvest handling will enjoy higher marginal returns (Unnevehr and Hirschhorn, 2000). Even in respect of end-markets that are not so demanding, higher volumes can be used to compensate for lower margins. Where markets are bifurcating between standard-intensive and bulk/anonymous segments, very large operators can pool inputs, labour, processing and handling facilities to enjoy economies of scale in both segments. It is such operations, in regions such as Brazil's Cerrado, that now dominate global agro-commodity production for soybeans, corn, beef, poultry, sugar and coffee.

Meanwhile, as a result of corporate financialization and accelerated capital accumulation and concentration, the value chains for virtually all agro-commodities are becoming global as well as 'shorter' (passing through fewer stages) and more concentrated in terms of their number of intermediaries and suppliers. Since African commodity-dependent countries typically have limited and resource-poor smallholder supply bases, they tend to be by-passed where supply-base concentration is pursued strategically. The overall result of these two processes is market marginalization: the increasing restriction of African countries, and of most producers in them, to residual lower-priced markets within OECD countries and – the data suggests – to markets in developing countries that are not experiencing as dynamic an expansion as those in, for example, East Asia (Gibbon, 2006).

• *The case of cotton*
Cotton is illustrative. Most West and Central African economies depend highly on activities in the cotton sector. Cotton production for instance, accounts for 5 to 10 per cent of GDP in Benin, Burkina Faso, Chad and Mali (Fortucci, 2003). About two million farmers in West and Central Africa produce cotton, amounting to roughly 30 per cent of total export earnings and more than 60 per cent of total agricultural exports. All these countries are low cost producers as opposed to the cotton producers in the developed countries, especially in the United States.

Drawing attention to the powerful vested interests in global commodity markets, a group of poor cotton-producing African countries have proposed that the developed countries eliminate all forms of cotton subsidies within a maximum period of four years (WTO, 2004). Their original demand at the Cancun WTO summit included a call for a sectoral initiative on cotton and financial compensation while subsidies were being removed. So far, the WTO reform on the cotton initiative has yielded little even though the WTO Agreement on Agriculture prohibits all forms of trade distorting domestic and exports subsidies. The lack of progress in the Doha round and the continuous trade distorting subsidies for cotton farmers in some of the world's wealthiest developed countries prompted Brazil, supported by two of West Africa's poorest countries (Benin and Chad), to pursue their grievances in the WTO Dispute Settlement Body (DSB).[8] In their submission, Benin and Chad emphatically noted that: '[t]he extraordinarily damaging impact of US subsidies in Africa has compelled Benin and Chad to participate in this case. For Benin and Chad and indeed for many of the least-developed countries of Africa – this appeal is unquestionably the most important dispute ever brought to the WTO.'[9] The Appellate Body ruled in

---

8. Appellate Body Report, United States – Subsidies on Upland Cotton, WT/DS267/AB/R, adopted on March 21, 2005.

9. Third Party Submission of Benin and Chad to the WTO Appellate Body, 16 November 2004.

their favour in a landmark judgment that will be remembered as a definitive piece of jurisprudence in the area of agricultural trade.

• *Future prospects*
In addition to the opening for African cotton producers arising from this legal landmark, there are still economic opportunities for smaller, poorer countries and small-scale producers to participate in some of the global commodity markets. This is partly because of the ongoing salience of varietal, area of origin and traditional 'good quality' differentiation in some segments of the markets for some tropical products. Historically, family farming systems have been superior in assuring these dimensions because of their comparative advantage in monitoring labour. Similar considerations apply to certain new quality dimensions emphasizing 'sustainability'. Demand may also expand in Asia for crops where quality demands are less exacting, but where family farming also predominates (e.g. cocoa). Furthermore, smaller producers can compensate for lower scale economies on the basis of becoming organized in larger entities.

Some crops, such as fresh vegetables and cut flowers, are not well-suited to the flat, empty landscape of the Brazilian Cerrado. Instead they require large volumes of labour and water, allowing highland tropical regions in Africa and elsewhere to be competitive in OECD markets, when these conditions are complemented by access to capital for investment in core large-scale farm operations and by good infrastructure. Secondly, again because of their labour surpluses, African countries are not at a disadvantage in most strictly presentational forms of product differentiation favoured in OECD markets (e.g. washing, slicing and mixing for salad, heads-off shrimp, fish fillets, etc) since these are typically labour intensive. Thirdly the countries with family farming systems should have the capacity to produce to those new quality standards where monitoring labour is critical to conformity. The main preconditions for them to improve their position overall are that they attain greater economies of scale. Better access to inputs, finance and extension, and more effective national systems for institutionalizing traditional quality dimensions would also generate considerable potential benefits. Further prerequisites for improved agro-commodity performance include meeting initial costs of conformity to those new quality dimensions dependent on monitoring labour, and meeting costs for improvements to trade-related infrastructure.

The recent surge in the world demand for bio-fuels has led to a new interest in agricultural land, this time not for food but for fuel export to enable developed countries to reduce their emissions of green house gases from the transport sector. This has spurred a rapid increase in food prices and an emerging competition for land for fuel as opposed to food, which may generate export incomes but increase food insecurity in developing countries and impact on smallholders' land access and labour viability.[10]

In 2001, the global trade on agriculture and agro-industrial products was estimated at $547 million or 9.1 per cent of total merchandise trade. A year later, similar studies showed a marginal increase to $583 billion or 9.3 per cent of total merchandise trade. Although the twentieth century witnessed agricultural trade increase in absolute terms, its share in comparison to world trade has decreased and continues to plummet. For instance, as a component of merchandise trade, between 1980 and 1997, agricultural trade declined between 17 per cent and 10 per cent. Yet, trade in agriculture remains the backbone of the economies of most Sub-Saharan African countries and other small developing countries.

Much will depend on how far African producers can begin to produce on a more level playing field relative to the highly capitalized, subsidized and sometimes tariff-protected farmers in key developed countries. In 2001, the World Trade Organization trade and finance ministers called for developed countries to put development at the forefront of the agricultural negotiations by taking measures such as decoupling domestic and export subsidies on sugar, beef, citrus, grain, cotton and other agricultural products. Succeeding multilateral trade rounds have been shaped by the debates on ag-

---

10. A recent OECD/FAO investigation has shown that "increased demand for bio-fuels is causing fundamental changes to agricultural markets that could drive up world prices for many farm products." In Brazil annual ethanol production is projected to reach 44 billion litres by 2016 compared to around 21 billion litres today (FAO, 2007). Bio-fuel production is now spreading in many Sub-Saharan African countries, including Benin, Tanzania, Uganda and Zambia (African Biodiversity Network, July 2007). Due to irrigation needs, bio-fuel production from sugar cane competes for the best land with food production. Brazilian and Latin-American experiences from the cultivation, harvesting (manually) and processing of sugar cane into ethanol have shown grave environmental and health implications (Comar and Gusman Ferraz, 2007; Altieri and Bravo, 2007).

riculture. The December 2005 Hong Kong ministerial conference was no different. As was the case in Cancun in September 2003, the Hong Kong trade talks could not move forward as a consequence of disagreement among members on agricultural reforms. In July 2006 the Doha round was suspended and only slow progress has been recorded since the resumption of talks in January 2007.

Land Rights, Markets and Capital

Land issues and land tenure reform

Africa is characterised by a range of farming systems all with varied rights under multiple forms of tenure including: private landholding with freehold title deeds, communal public lands under customary tenure, and state-held land where the state retains legal ownership upon which various forms of tenure based on either leaseholds or permit systems have been devised by the state, underpinned by complex legal and administrative systems. Usually the state bureaucracy plays a significant role in rural land administration, with traditional leaders being provided with limited responsibilities over land management and people in areas where usufruct rights to the land are still practised. The household and individual plots and commons found on customary lands provide subsistence to millions of people. Nonetheless, the implementation of market liberalization and democratization policies has had an indirect if not direct impact on customary management arrangements. The introduction of modern forms of governance based on elections and statutory arrangements has, in some cases, been the beginning of dysfunctional combinations of old and new institutions and practices (Adams et al., 1999).

This situation has made rural land ownership a key issue in Africa constituting a problem that has largely remained unresolved in many countries since colonial times (Rukuni et al., 2006). Historical conflicts and inequities over access and ownership of land rooted in colonial land dispossession are intense in countries such as South Africa, Namibia, Malawi, Kenya and Zimbabwe. Clearly, inequitable land distribution in Africa relates also to rural poverty and political instability. Increasing tensions over land are found in the ethnic violence in northern Ghana (1994–95), the land violence in the Tana River district of Kenya (2001), the civil war in Rwanda, civil eruption in the Ivory Coast and Zimbabwe's land occupations and violence.

Land rights, land distribution, use and management are intimately linked to the performance of African agriculture. Problems of land insecurity and conflicting claims on land form important reasons for Africa's low agricultural productivity. Access to land is indispensable for survival, given that the struggles for land in Africa are not confined to peasants. The social base and leadership of the 'land hungry' include the landless, farm workers, retrenched mineworkers, industrial and urban-based workers, a diverse strata of the 'semi-proletariat', even some middle income and rich peasants, and the middle classes. Where tenure reforms have been introduced, local farmers are often uncertain about the nature of their rights and confused about the extent to which institutions and laws affect them. Matters are further clouded by local and national political conflicts over land management roles in areas where traditional customary land law prevails. The development and implementation of effective land policies, including clarifications on land tenure systems, are crucial aspects of economic governance within the smallholder sector. Local institutions are usually disempowered and weakened with little role in the regulations and enforcement of sanctions.

As *de jure* and *de facto* land reform processes unfold in African countries, they do so, more often than not, in a piecemeal fashion, disconnected from a wider development strategy, which is likely to generate limited returns or falter. Almost all state-led land reform has the declared objectives of poverty reduction, equity, employment creation and land tenure change. The South African land reform programme, for example, sought to 'reduce overcrowding in the former homeland areas and expand opportunities for rural people, and to improve nutrition and income for those who choose to farm among rural households' (Ntsebeza, 2004 and Cousins and Claassens, 2006). The initial Zimbabwe reform programme aimed to 'reduce poverty among rural households and farm workers, and to achieve domestic food self-sufficiency' (Matondi, 2001). Land redistribution in northern Ethiopia aimed at securing land for everyone, however, in the process it made many farms too small to be economically viable (Atakilte Beyene, 2003). Laudable as these objectives may be, they have not been integrated into comprehensive development strategies that link land redistribution with provision of the necessary infrastructure, services and inputs to enhance land productivity and provide viable farm-based sustainable livelihoods.

Meanwhile, the ongoing effect of HIV/AIDS is impacting on rural land usage and tenure patterns. HIV/AIDS is experienced unevenly across space. Southern Africa, particularly South Africa, Swaziland, Botswana and Zimbabwe, are the most affected by the pandemic. Poor farmer households are the most vulnerable, often forfeiting or alienating their land rights as a result of sickness or death within their families and households. The most marginal rural households (both male- and female-headed) are likely to break up and disappear altogether. The pandemic is encouraging shifts to new forms of tenure, e.g. rental or increased land sales, as well as new patterns of cropping and land use (Izumi, 2006 and 2007). Furthermore, the pandemic is bringing the negative impact of aspects of customary law on the livelihoods of women and children into increasingly sharp focus. Across Africa, the land rights of women and children in patrilineal areas are becoming ever more vulnerable to dispossession by kin in patrilineal areas on the death of male household heads.

Land individualisation and markets

Historically land markets have been proposed and implemented with mixed results in various African countries. The provision of individual legal titles, where smallholder plots are demarcated and provided with complex certificates of occupation, was implemented in Kenya and Zimbabwe before national independence. However, this titling process largely failed as it existed in a state of limbo between the larger customary system of land rights regarded tacitly as state land, and the large scale freehold land usually owned by a preserve of settler farmers and foreign investors.

During the 1970s and 1980s, titling and privatization of rural land in Africa entered the World Bank's agenda. African farmers' values were perceived to contradict market operations. The inference was that titling and privatization of land would break tradition-bound social values and induce a land market. In a significant departure from the World Bank's otherwise consistent efforts to promote the extension of market relations throughout African commodity, labour and capital exchange the World Bank now stresses that a rural pro-poor agenda requires attention to customary tenure rights and land management systems (World Bank, 2003). The World Bank position is now supportive of evolutionary land tenure, seeing customary tenure as central for ensuring the poor's security as local tenure regimes evolve towards market-based practices (Bruce and Mighot-Adholla, 1994, Platteau, 1996). To stave criticism that it is supporting traditionalism, the

World Bank has tried to press for reforms of the traditional authorities safeguarding customary land tenure, and in so doing asserts that customary land tenure can strengthen women's land rights, promote decentralized land institutions, and raise productivity – features rarely if ever identified with customary tenure in the past (WDR 2008 May version p. 6.2).

Large versus small scale farmers – Growing landlessness

The reality is that customary land rights are no longer the central issue in many African countries. Smallholder farmers are often in competition with large-scale farmers who receive preferential state support despite strong evidence that smallholder farmers are more equitable and more efficient per unit of land. Small farmers have already been or are currently being pushed into vulnerable ecological areas outside their traditional home areas.

Although more than half of the arable land that is idle in the world is in Africa, the land area in some countries is near maximum population density given the present agricultural technology and lack of soil fertilization. Africa, not normally associated with landlessness, is now witnessing growing numbers of vulnerable landless people particularly in South Africa, Zimbabwe, Kenya, Ivory Coast and Namibia. The landless are the product of intensifying demographic pressures in rural areas, retrenchment or eviction of powerless minorities and farm workers from farms and plantations and marginalization of pastoralists.

Subsistence food needs are forcing smallholder farmers to expand the area under cultivation into forests and more and more marginal and fragile areas characterised by poor rainfall, degraded soils, and deforested lands in contrast to large-scale farmers and foreigners who manage to procure land of high economic value. In some cases, large-scale farmers own large tracts of land that are underutilized at the same time as small farmers are struggling to farm their small plots with poor soils. This is evident, for example, in Rwanda's congested rural areas. Local people resort to cultivating steep slopes with inadequate conservation skills for managing soils and water resources.

At the intra-household level, youths and women tend to be marginalized relative to men in accessing and using land, which has compounded poverty in the smallholder sector (Manji, 2002, Peters, 2002). Inter-household land conflicts of a multi-dimensional nature arise as well, related to the unequal local distribution of resources, the politicisation of ethnic groups, the manipulation of religious differences and social exclusion (Jua and Nkwisi, 2001). In other words, smallholder farming households' ability to sustain the soils and their farms is deeply compromised by the lack of access to economic or other resources that affect their security of tenure.

In general, African governments, pressurized to encourage private enterprise and foreign investment, have lacked the political will to challenge the production base underpinned by large-scale commercial farmers because of their supposed superiority of production and diversified and complex portfolio of production at the farm level.[1] Large-scale farming tends to receive unqualified state support and protection of land rights against smallholders' demands for land redistribution. By contrast, smallholders on customary tenure lands, no longer have sufficient land nor even secure land rights to the land that they currently farm.[2]

Foreign land investments

African states have initiated a host of incentives (tax rebates, physical and moral security) for foreign investors to attract much-needed foreign currency into the country. Historically, the majority of investors and European settler farmers were concentrated in Southern Africa producing commercial export crops as well as food products such maize, wheat and beef (Selby, 2006). More recently, they have ventured into horticulture, safari ranching and tourism. Through their ownership of land, they wield significant economic power and privilege to this day. Generally, the external investor is unaffected by the rigours of local policymaking or the rules and regulations under which smallholder agriculture is controlled.

Besides encouragement of foreign investment in the agricultural sector, there are new extensive land uses in the wildlife and tourism sector, diversification of crop production including a rapidly growing interest in biofuel, and creation of export processing zones in the agricultural sectors have

---

1. 'Lack of political will' masks a more complex political context in relation to the state and resource control, e.g. where factions of the government in various ways also gain from the privatisation and liberalisation processes.

2. The developments in Zimbabwe since 2000 are an exception where the state has expropriated more than one thousand large-scale farms, mainly in the hands of white settlers and companies, without compensation.

been encouraged. Increasing demand for resources for both local and export markets as well as escalating competition for the control of natural resources have been a source of insecurity and have increased the incidence of land and natural resource conflicts. Land conflicts are deeper in relation to land with valuable resources such as minerals (notably oil, diamonds, gold, and other precious metals and minerals), ecologically suitable land (with good soils, water sources, densely forested), and with high potential for crop production, and well-developed infrastructure developments (communication, dams, irrigation).

In post-conflict countries recovering from civil wars such as Angola, Mozambique, Rwanda, Sierra Leone, and Liberia local elites and foreign investors seek to secure rights over the best land, close to markets through international legally protected agreements specifying land purchase and lease terms that erode rights to customary lands and common property resources and undermine access of returning refugees (Norfolk and Soberano, 2000, Kairaba, 2002).

Africa's uncapitalized smallholder production: Three decades of declining investment and agricultural extension

In stark contrast to Asia, Africa remains seriously food insecure. The investment in improved agricultural input packages and extension support tapered off and eventually disappeared in most rural areas of Africa under SAP. Concern for boosting smallholders' productivity was abandoned. Not only were governments rolled back, foreign aid to agriculture dwindled.[3] African governments could no longer afford to offer cheap agricultural credit and the private sector did not give loans for staple food crops. Private credit institutions were highly selective and only reached a minority of farmers. Smallholder peasants on communal land without private land-titles lacked collateral for obtaining formal agricultural credit.

Likewise, with the rolling back of the state, extension services have virtually collapsed. The much hoped for solution is 'demand-driven extension' which is to be supplied by the private sector and NGOs as advocated in the WDR 2008. So far, NGOs have not provided anywhere near sufficient extension

---

3. World Bank funding for agriculture itself declined markedly from 32 per cent of total lending in 1976–8 to 11.7 per cent in 1997–9 (Pincus, 2001:196).

services. At a critical juncture when African smallholders need to be making timely investments and introducing agricultural innovation, signs of agricultural disinvestment are becoming ever more common.

In many densely populated areas, farm plots are shrinking and the habit of letting land lie fallow is increasingly being abandoned. Instead it is intensively farmed, usually for maize or other basic subsistence crops but without fertilizer, which many poor peasant households can no longer afford. This leads to soil-mining, decreasing yields and aggravated poverty. Farmers' prevailing level of skills and resources are not enough to handle such situations, although some positive shifts have been undertaken to develop more systematically low-input and organic agriculture.

The CGIARs and national agricultural research institutes have developed a range of new and improved environment-friendly methods for soil and water management as well as a large number of crops and crop varieties that are adapted to many of Sub-Saharan Africa's variegated ecologies and farming systems. Together they represent a real possibility to turn around the downward trend in food security and enhancement of agricultural productivity. To disseminate them would, however, demand a massive investment in agricultural extension services towards an African sustainable agricultural revolution along the lines of the green revolution in South Asia, which the WDR 2008 has dismissed as not feasible in Africa's environmental context.

Meanwhile, climate change is having a pronounced impact in Africa's semi-arid areas and the vast zone around the equator. Whereas the exact effects are still unknown, it is quite clear that Sub-Saharan Africa will need to make far-reaching adaptations in its farming systems to accommodate changed rainfall patterns and cropping seasons, by planting new crops and crop varieties and adopting new farming practices. Africa's vulnerability is exacerbated by its heavy dependency on the narrow range of agricultural products to support its economies, which often fail due to pest outbreaks, climate variation, price fluctuations etc. Climatic variability and change, inappropriate land use or land tenure policies, add to the environmental pressures that result in further food insecurity for rural people. This again calls for more, not less, research and extension.

The HIV/AIDS pandemic is having a seriously draining effect on smallholder farming labour supply. Many countries now face losing a third of their agricultural labour force due to the disease. Child-headed households are becoming increasingly common. Demand-led extension presupposes a relatively experienced farmer who knows what to demand and who can prioritise and fit new advice to an already existing body of knowledge. Child-headed households risk being by-passed by existing extension providers.

So far, market based solutions have not been able to respond to the rising need for increased and more varied extension services. The history of state-led extension in Sub-Saharan Africa is full of stories about ineffective services. Under-resourced state extension personnel had to become even more selective and now often only provide extension services to a minority of 'progressive farmers' located in easily accessible rural areas. Some NGOs administer extension services but because NGOs are unevenly spread over the countryside, their services tend to be patchy. Much NGO literature advocates demand-led extension, which in reality is seldom on offer. Most NGOs do not have the capacity to respond to the expressed demands of smallholder farmers and they have knowledge and a pre-determined preference for certain practices and/or technologies while ignoring others. Hence, their supply is limited. Much NGO administered extension is also coupled to projects, which to varying degrees are financed by the NGO and international donors. In order to obtain extension advice or projects, potential beneficiaries learn to demand what the NGO is able to supply – what the World Bank calls 'supply-driven demand-driven' extension. Recent and broader initiatives and support related to these areas have also been taken by the United Nations Food and Agriculture Organisation (FAO) and the International Fund for Agricultural Development (IFAD), e.g. to the so-called Farmer Field School Movement (IFAD, 2002, Duveskog, 2007). The scaling up and sustainability of such initiatives are, however, still uncertain.

A far more broad-based solution to smallholder research and extension needs is required. In particular, the need for mutual learning related to approaches, practices and technologies for sustainable agriculture is increasing. In many cases the situation calls for *new* solutions in the form of a comprehensive African sustainable agricultural revolution to be promoted by smallholders, smallholder groups, extension workers, the private sector, researchers and donors in close cooperation. In such a context, community networking, farmer-to-farmer extension and formation of smallholder farmers groups or various local organisations can only be part of the solution.

Institutional Supports for African Smallholder Agriculture

African nation-states

Their role in agrarian development

Emerging from current research and policy debates on the roles of African states in promoting agriculture are a number of tasks that nation-states are supposed to fulfil, namely: policy formulation, coordination and evaluation, provisioning of public goods, protection of property rights, regulation, coordination and overcoming market failures, reducing vulnerability and inequity.

Specifically with respect to agricultural marketing, 'getting institutions right' according to Gabre-Madhin (2006) includes: 1) mechanisms to transparently grade and standardize products for market, from the production level on throughout the market chain, 2) market information that is accessible to all market actors, 3) fostering competitive practices among all market actors, across all levels of the chain, 4) financial markets to respond to market needs for trade finance, for inventory finance, and for alternative financial products, 5) dispute settlement and regulatory systems to evolve according to market needs, and in a way that also relies on the private incentives for self-regulation, notably through the potential role of trade associations, 6) risk-transfer through mechanisms such as forward contracts and transferable warehouse receipts, and, 7) concerted efforts to build capacity throughout the marketing system, including cooperatives, small and medium private traders, and public actors.

An underlying assumption is that the state should provide for the rule of law, and a functioning judicial system more broadly. It should support producer organizations by providing a regulatory framework and encouraging their active role in relation to various input and product market functions. Lately, a debate has started about whether the state should provide subsidized fertilizer in order to help increase crop productivity. To support productivity more generally, the state should provide infrastructure, and play an important role in providing basic social services such as health care and education. In all this, it needs to function efficiently and adhere to the values of transparency and accountability. Legitimacy stems from the state being able to serve, or work, in the interest of its citizens but even the most legitimate state may lose its legitimacy if it is unable to deliver in an efficient way (Lipset, 1984, Rothstein, 1994). But to what extent are African states able to live up to this?

Disputed character and functioning of African states

Over the past 25 years, the World Bank has portrayed the African state as the major institutional constraint blocking rural development. The WDR 2008 continues in this long tradition. Mainstream western political science research portrays a colourful array of adjectives to describe the African post-colonial state including: 'pretender', 'parasitic', 'personality', 'clientelist', 'kleptocratic', 'unsteady', 'over-extended', 'predatory', 'crony', 'soft', 'weak', 'lame', 'rentier', 'sultanist', and finally 'neo-patrimonial' (Olukoshi, 1998:14). Neo-patrimonialism refers to states controlled by a 'Big Man' whose personalized authority is combined in one way or another with westernized laws and systems (Bratton and van de Walle, 1997, Clapham, 1985). The role of the Big Man is to provide his clients with material goods, and it is from this process of redistribution that his legitimacy derives (Chabal and Daloz, 1999).

This theoretical framework claims that there is no actual pressure to reform such a system based on patron-client relationships and that civil society is governed by the same principle of centralization-redistribution, of patron-client relations. Further, there are not many opponents to the 'Big Man' available, since it takes a broad platform of clients to ensure a power base broad enough to challenge an incumbent ruler. This may also explain the surprising continuity in African leadership, where formerly authoritarian rulers may reappear as democratically elected leaders. Chabal and Daloz (1999) argue that African politics differs radically from western models with respect to: 1) the boundaries of politics between the private and the public, 2) the notion of the individual as part of a collective unit, 3) the source of the legitimacy of power, 4) different forms and functions of representation as compared with the West, and 5) the meaning and role of political opposition, which complements rather than opposes the ruling circles. The African state has not been sufficiently separated from society[1] and pre-colonial patterns of societal relations still influence the way the state works.

Mamdani (1996) argues to the contrary, stressing the colonial legacy of a bifurcated state encompassing 'citizens' (urbanites under westernized law) and 'subjects' (rural dwellers under customary law). In the rural setting, the fusion of various spheres of power in local chiefs made these the ultimate tool for indirect colonial rule. Since decolonization, this fused power in the hands of chiefs lingers on, and local chiefs in many countries retain an unprecedented power base, which is instrumental in upholding their positions, which they often use to establish themselves as patrons.[2]

However, western donors increasingly acknowledge that African states require more policy autonomy, hastening to add that this must be accompanied by enhanced legitimacy and efficiency. It is vital that the state not only is able to take on the vast and increasing number of roles that are ascribed to it, but also that its capacity to fulfil these roles efficiently increases. Both the scope and the strength of states need to increase. Reforms driven by internal forces – democratic institutions, social movements and domestic leaders – are necessary to move in the direction of autonomy and efficiency without decreasing legitimacy (Mkandawire, 1996).

Resource mobilization capacity of the state

African states have to mobilize financial resources that involve external flows, internal flows directly generated by the state, and internal flows generated by other actors than the state, under the influence of government action. External flows encompass aid, loans, remittances, trade flows and foreign investments (equity and portfolio). From a Sub-Saharan African

---

1. The concept they use is the 'institutional emancipation' of the state from society (Chabal and Daloz 1999:13). However, their use of 'institutional' in this sense is misleading, since they do not use the concept as broadly as in the tradition following North (North and Thomas, 1970 and 1971, North, 1993).

2. The retention of power in local traditional chiefs, or in some cases, the process of re-traditionalisation of African societies, also has to be seen in the contemporary processes of state building and party politics.

perspective, aid and loans are the most significant. However, there are several problems related to external flows: first, a long-term growth strategy based on external borrowing is rarely sustainable. Second, the extent and conditions under which foreign aid is able to promote economic growth need to be considered (Easterly, 2003, Gunning, 2004). Third, states can and do become institutionally dependent on aid making them vulnerable to external forces (Elbadawi and Gelb, 2003). Fourth, the international development architecture may be unable to effectively and efficiently provide relevant forms of finance for development (Sagasti, et al., 2005).

Problems with external resources point to the central importance of African states in generating domestic resources. State capacity to directly generate such resources basically concerns the 'tax effort', which is the ratio of actual tax revenue to some measure of 'taxable capacity' or tax base. There is also the question of what capacity the state has to induce other domestic actors to generate resources, that is, what capacity it has to generate economic growth. In that sense it may be argued that the quality of the agricultural policies of African states is a necessary, however not sufficient, precondition for the possibilities of African states to support agricultural development.[3] This would then include all the various factors needed to foster agricultural development in the list of state roles above.

The restructuring and strengthening of the African state is a necessary pre-condition for an agricultural and rural agenda that is genuine in its support for smallholder agriculture and the promotion of an African sustainable agricultural revolution.

In the WDR 2008's own words, "comprehensive multisectoral approaches are required to coordinate the contributions of agriculture with investments in other sectors, raising complex issues of investment priorities, political tradeoffs in budgetary processes, and intersectoral coordination of implementation" (WDR 2008 ch. 10. pages 324–5, July version). At the same time the new development assistance architecture comprises budget support, sector based strategies, basket funding, national dialogues and recipient ownership. This all points in the direction of increasing transpar-

---

3. There exist a number of policies related to African agricultural development that are seemingly sound, but they are not, or are only partly, implemented. The gaps between policies and their implementation often reflect a lack of understanding of policy makers of the politics of actual development practices that may promote or hinder policies chosen for implementation.

ency, accountability and trust in the state as planner and implementer of national and agricultural development.

There is, however, a glaring discrepancy in World Bank and donor funding trends. On the one hand the African state is conceptualized as an obstruction to development and, on the other hand, since the Paris Declaration, development assistance has shifted from project and programme aid to budget support, which necessarily relies heavily on state agencies' efficiency and accountability. However, the World Bank seems to have a reluctance to give up its skepticism about the African state and continues to fall back on the problems of poor state governance in cases of policy failure. In other words, poor policy performance can also be seen more as a feature of the World Bank's lack of accountability over the last 25 years than a feature of the African states who have only had limited scope to define their own development policies (Wangwe, 1987 and Fundanga, 1996).

Rural households as transforming institutions interacting with market and state trajectories

Peasant family farming, stretching over vast swathes of the continent, has accounted for the largest numbers of agricultural producers in Africa over the past century. While markets and states determine the policy framework within which rural change takes place, individual smallholder peasant household agency drives the trajectories of rural development or material impoverishment witnessed over the last century. Therefore it is critical to understand the specific conditions in which rural households operate for any agricultural development or poverty alleviation strategy to be effective.

To understand the scope for rural smallholder farmers' decision-making, it is useful to outline the classic characteristics of peasant farming. Peasant smallholders are identified with the pursuit of an agricultural livelihood, which combines subsistence and commodity production. Their internal social organization revolves around the family as the basic unit of production, consumption, reproduction, socialization, welfare and risk-spreading. Externally subordinated to state authorities and regional or international markets that involve transfers of tax and profit, peasants are part of a wider system of class differentiation. Residing in rural settlements with widely dispersed households or nucleated villages, they tended historically to form communities espousing a traditionalist outlook, led by tribal or other local authority figures who have had responsibility for enforcing order and adherence to community values both under indirect colonial and post-colonial rule (Bryceson, 2000).

The WDR 2008 resorts to stereotyping African smallholders as 'subsistence producers'.[4] This is far from the case. The history of African rural economies and politics testifies to this. During the decades of European colonialism spanning the 1890s and the first half of the 20th century, African farming households, spurred by colonial taxation, increasingly congealed as peasantries producing agricultural commodities in the form of the 'beverage crops' – coffee, cocoa and tea – as well as cotton, tobacco, groundnuts and cashewnuts. These became the continent's major exports. Colonial governments utilized various traditional native authority models to govern, which tended to foster a traditionalist outlook. However, following World War II, African nationalism gained ascendancy. Mass political support from the peasantry in one country after another catapulted African nationalist leaders into the driving seats of independent nation-states. Thereafter most African post-colonial governments pursued policies aimed at extending, capitalizing and modernizing peasant production combined with education, health and service supply in order to raise peasant productivity and living standards as a foundation for industrialization efforts. This was the development trajectory that prevailed until the economic upheaval of the international oil crisis followed by SAP and economic liberalization policies.

Peasants' adjustments to increasing agricultural input costs and poor market prospects have in many cases led to a reallocation of land and labour away from commercial agriculture. In the wake of the oil crises of the mid and late 1970s, widely geographically-dispersed, under-capitalized African peasants found it difficult to produce their traditional cash-crops competitively in the world market. Large-scale Asian production of various tropical crops arising from earlier investment was coming on-stream. African peasant agricultural commodity production was increasingly losing its place in the world division of labour. At the same time as returns from peasants' com-

---

4. Interestingly, the WDR 1982 report on agriculture adopted a more upbeat position: 'Far from being "tradition-bound peasants", farmers have shown that they share a rationality that far outweighs differences in their social and ecological conditions' (WDR 1982:91).

mercial agriculture became less certain, daily cash requirements increased under the economic stringency of SAPs. Subsidies in agriculture, education and health were removed. School fees and 'user fees' at health centres became a high priority in peasant household budgets. Market liberalization from the perspective of the rural consumer, tended to expand choice but at arm's length, for much of the tantalizing merchandise came with unaffordable prices.

African farming households were highly responsive to these changes in their terms of trade and public service provisioning. A 'scramble in Africa' ensued in which the search for viable livelihood alternatives was paramount (Bryceson, 2002a). There were a number of discernible economic trends, which are described in the next section. Concurrently, associated social trends were unfolding but to understand the significance of them it is necessary to sketch the general pattern and in particular the gender relationships that prevailed before the scramble.

Traditionally, the head of a household held a privileged position. Where men-headed households are predominant, their privileges do not always translate into household welfare.[5] The unequal resource access has mirrored the gendered division of roles and responsibilities in rural households. Although these are location-specific, women have generally been responsible for the home-based activities 'inside', and often had limited mobility outside, while men usually had the primary responsibility for the agricultural tasks related to the 'outside'. However, in reality rural women have tended to contribute a great deal and often the largest share of farming effort within the household, yet their role has been largely invisible due to cultural norms viewing the man as the 'farmer'. Historically, a gendered division of labour has permeated African farming cycles, which made women responsible for some crops (often staple foods), and men for others (often cash-crops) and livestock.

Clearly, in situations when men worked in towns but retained the responsibility to make agricultural decisions, this division led to agricultural inefficiencies and slowed down poverty alleviation efforts. For example,

---

5. Traditionally men have headed most African rural households. However, the development of migrant labour systems, in particular in Southern Africa, left many households with female heads. The structure of households is changing rapidly in the current context due to the impact of HIV/AIDS and intensified migration to urban areas, thus increasing the number of female- as well as child-headed households.

important agricultural decisions may be delayed, such as regarding what to plant and when. Women and men also have had different access to local markets, and this affected the impact of trade liberalization on the most vulnerable. Indeed, while Africa has in general been slow to move up the agricultural value chain, women have been less able to do so than men. In Ethiopia, for example, a strict line has existed between products traded by women and those traded by men. Women have traditionally been responsible for the trade of low-value items in the informal segments of the market places, while men traded in the formal segments of the market (Torkelsson, forthcoming). It is for this reason that female heads of households have faced a triply disadvantaged position because without a male spouse, their households have an inferior land and labour position as well as limited non-farm options.

Rural women tend to be the 'poorest of the poor'. They eat 'least and last', yet they are responsible for ensuring the food security of children. Rural women's access to resources is limited across the range of rural resources. They have access to less land, education, extension services, financial inputs, and other important agricultural inputs, such as fertilizers compared to men. Moreover, rural women have limited productive associational ties, and less time to participate in local social networks.

In the scramble for alternative income to make up for the decline in African male heads of household's cash-crop income-earning, women and male youth have been very active. Their rising income at the same time as the male head's cash earnings have declined affects the gender/generational power balance within rural households. The clear intra-familial division of labour in which women are the subsistence food producers and men are the cash-crop producers has started breaking down. Male heads of households no longer monopolize the family's cash earnings. Both the gender and generational division of labour have blurred as the cash economy slips from being the preserve of male household heads.

Individualization of economic activity has had a dissolving effect not only on the long-standing agrarian division of labour, but has also eroded the sense of economic rights and responsibilities within rural households. Pooling of income within the domestic unit has weakened as categories of people who formerly were not expected to earn income now simultaneously receive less from male heads of household, and assert their right to determine how their income is going to be spent. At the same time, the impact of HIV/

AIDS on the composition of the productive population, leading to new forms of households, such as those headed by children or the elderly, has altered both the composition and form of African rural households.

African Rural Agency in Response to Global Market Pressures

The preceding sections have provided evidence of the eroding position African rural households have faced in international agricultural commodity markets since the 1980s. The continental-wide exodus of smallholder peasant farmers from export crop production lies behind the exceptionally rapid displacement of agricultural labour and has set in train a process of deagrarianization involving occupational adjustment, income-earning reorientation, social identification and spatial relocation of rural dwellers away from strictly agricultural-based modes of livelihood. After a century of conscious colonial and post-colonial policy support to the formation and maintenance of peasant household production units, the absence of supportive policies to mediate the full force of changing market forces has triggered depeasantization.[1] Depeasantization is a specific form of deagrarianization in which smallholder farming households lose their economic capacity and social coherence, demographically shrink in size, and unravel as peasant communities (Bryceson, 2002a). Profound economic and cultural transformation in rural areas has take place through depeasantization. Labour expenditure is less household-based and agriculturally focused. Rural intra-household relations are now characterized by more individualized decision-making. Local social norms are breaking down and inter-household economic differentiation is generating winners and losers who undermine the egalitarian legacy of tribal communities.

---

1. Although the process of depeasantization is not uniform, since it can be diverse across the continent and within countries, the term nevertheless captures the major direction of trends unfolding in African rural areas today. This does not preclude that processes of repeasantization may occur in specific areas or countries, such as Zimbabwe, but this does not represent the direction of current general trends on the continent. Neither does it imply that the process of diversification of peasant labour and migration is a recent phenomenon, but rather that the process is intensifying and broadening.

Depeasantization consists of a bundle of inter-connected economic, social and political trends. First, there has been a surge of a variety of non-agricultural activities, notably trade and mining, in place of export crop production. Second, increasing cash-crop production has displaced community exchange labour. Thus activities such as beer brewing, midwifery, hair plaiting, and local entertainment, which formerly were done on some sort of local exchange basis or as a contribution to village life are increasingly more apt to be performed for cash. Third, money-earning on the part of various categories of family labour has become more common. Women, as wives, as well as youth and even in some cases children, have joined male heads of household in working for cash as outlined above. Fourth, households have gained multiple income streams, which are not always pooled within the household. Wives and youth, in many cases, have acquired some degree of autonomy over their earnings and made their own discretionary purchases with their money.

Fifth, work experimentation is widely prevalent. Engagement in non-agricultural activities is no longer reserved for the agricultural off-season. Individuals might pursue two or more livelihood activities simultaneously or serially switch from one activity to another in the process of experimentation, trying to offset losses in one with gains in another. Sixth, in many areas rural local purchasing power has imposed severe constraints such that people, especially youth, are motivated to be more mobile or migrate in order to facilitate their trading or other occupational activities. Seventh, while it has been observed that cash-based work is on the increase, people have retained a strong desire to achieve household basic staple food self-sufficiency as much as possible to avert having to be reliant on purchased food. The issue is to achieve a balance between subsistence food and cash production such that the household always has a subsistence fallback in case any or all of their other cash-earning activities fail.

The following examines and concretises the framework and mechanisms of depeasantization with respect to the three main escape routes from smallholding peasant farming that the WRD 2008 proposes, showing that smallholder farmers, have already become active agents in these spheres.

Non-agricultural rural income diversification

There has been a surge of non-farm production, revealed through local-level household surveys, which has measured the proportion of non-farm production on average at between 40 to 50 per cent (Reardon, 1997, Ellis and Freeman, 2005). When subsistence agricultural production is omitted from the calculation, non-agricultural household income is generally between 60 and 80 percent of total household income earnings (Bryceson, 2002b).

Types of diversified non-agricultural activities vary from place to place, ranging from modern to traditional, high to low-income-earning, and formal to informal. They fall into four basic 'complexes' related to specific regional agro-economic zones: first, local services which dominate in remote areas consisting primarily of services as well as some handicraft activities catering to the restricted local market. Beer brewing and alcohol distilling are very common income-earning activities particularly for the poor. Trade, the second complex, has come to the fore and spreads ubiquitously in areas with histories of active participation in labour migration and/or agricultural commodity production. These are market-responsive areas with mobile local populations that are aware of income-earning opportunities and consumer demand beyond their immediate locality, often facilitated by historical links with urban areas.

The third complex consists of mining and other concentrated exploitation of natural resources by local populations and rural migrants, sometimes under 'boom' conditions and a 'get-rich-quick' mentality. The outstanding example of this is small-scale artisanal mining for gold, diamonds and other precious stones, which can be a major rural labour-absorbing activity in mineral rich countries such as Zimbabwe, Tanzania, Ghana, and Sierra Leone. Logging of hardwoods from Africa's forests is another example. The small-scale producers are operating informally and sometimes illegally in a highly competitive unregulated environment often tempered by a buccaneering as well as a desperate spirit shared amongst producers.

Fourthly, transfer payments are increasingly significant in agrarian marginal areas that have experienced heavy rural out-migration and the absence of peasant agricultural commodity production notably in parts of West and Southern Africa. In South Africa, for example, the rural male population has been extremely mobile, characterised by geographical movement between the countryside, on one hand, and towns or mines, on the other. In parts of rural Ghana and Senegal, international migration to Europe and North America has become prevalent. Psychological, cultural and material ties to their home areas remain. The remittances that they send back in effect represent rural non-agricultural earnings of a passive nature. In South Africa transfer payments, either pensions or remittances from non-resident relations, are in many rural areas the most dominant income stream for the local population.

Urban migration and employment

Income diversification has generated a great deal of spatial mobility, some of which is urban-focused. It needs to be noted that the character of urban migration has altered since its initial surge in the post-independence period. At that time, youth were attracted to the pull of capital cities and the contrast between tradition-bound rural areas and urban modernity was stark. Now, by contrast, as non-agricultural commodity trade, private services and labour markets expand in the rural areas, the economic and cultural gap between the two seems to be shrinking, especially in the perceptions of rural dwellers.

Urban migration is pronounced but it is no longer focused primarily on the capital cities. In most countries the urban primacy of the capital has dropped dramatically and secondary towns as well as fast-growing settlements associated with mining and the exploitation of hitherto rural-based resources like fish or timber have served as magnets for migrants eager to seize the economic opportunities they promise (Bryceson, 2006). Between 1975 and 2005 the urban population rose from 21 to 35 of the total population in Sub-Saharan Africa (World Development Indicators, 2007). But African cities are fragile economically, socially, environmentally and culturally. An interesting expression of this fragility is the widespread presence of urban farming, highly prevalent in smaller secondary cities where land is still relatively abundant, but also to be found in and around most of Africa's large capital cities. Many first-generation migrants still wish to pursue farming for cultural reasons, but above all, urban farming, which consists primarily of staple food crops in most cities, relates to people's need for a subsistence fallback in lean times.

Other urban ties to the land arise in relation to urban dwellers' practice of circular migration, rotating their residence between urban and rural areas to avail themselves of access to rural livelihoods and support networks. So too, many migrants choose to live in peri-urban areas where they can pursue both urban and rural livelihoods readily from their homes without having to commute between rural and urban occupational options. Until the foundations of African urban growth are based on the 'urban pull' of firm employment prospects rather than the 'rural push' of declining agrarian prospects, urban migration remains more of a threat than a solution to African development and poverty alleviation. The continuous growth of the urban informal sectors and spread of urban poverty are indications of this threat.

Relinquishing smallholder production autonomy: Contract farming and agricultural wage labour

The WDR 2008 infers that the lack of competitiveness of African smallholder commodity production will necessarily catapult many farmers into contract farming or agricultural wage employment.

For producers, the advantage of contracts over open market relations is that they reduce the transaction costs of marketing and input procurement, as well as conferring greater credit-worthiness. Contracts are also associated with obtaining higher prices over the long run (MacDonald et al., 2004, ADB, 2002). Traditionally, long-term agro-producer contracts in developing countries were confined to crops such as tea and sugar, where attaining export quality required that exporters invest in factories close to fields. In turn, factories could be run economically only with a consistent supply of raw material.

The wider relevance of contracting in an African context lies in its potential for increasing economies of scale and assuring quality. Although contracting in Africa was traditionally confined to sugar, tea, rubber and tobacco, more recently it has emerged for cotton, table fish, and fresh vegetables. In most of these cases this has been in a context of broader 'scaling up' and greater competitiveness.

Besides conferring scale advantages, contract farming's association with monopsony allows buyers to provide inputs on credit, in the knowledge that the latter can be recovered in the process of crop purchase. In addition, it creates economies of scale for some post-harvest processing (e.g., pulping, fermenting and washing for mild arabica coffee) and for subsequent sorting (or testing) and grading. Together, these economies of scale make it more likely that the harvested crop will be of good quality and that good post-harvest practice will be followed. This likelihood is strengthened where growers are paid a premium for products of good quality.

While contract farming has been widely advocated as an effective instrument to link smallholders to markets and input supplies and is endorsed by the WDR 2008 based on the above outlined argument, it is, however, very selective in its geographical outreach – usually restricted to locations near big cities or major roads. Socially, over time it tends to exclude small producers, and the crops grown are primarily export cash-crops rather than food staples. The overall impact of contract farming is thus likely to be much smaller than anticipated. Moreover, contract farming constitutes a top-down take-it-or-leave-it approach with limited technical transfer. Undoubtedly it can benefit some farmers, but it is not an omnibus solution to low productivity and food insecurity for the majority of African peasant farmers.

Similar arguments are made for the efficiency of large-scale farm and plantation production. The question however remains whether in relinquishing their autonomy smallholders gain in terms of income and security of employment. Smallholders' bargaining power in contract farming can be very limited particularly in relation to the increasing influence of supermarket value chains. Agricultural wage labourers tend to have even less room for manoeuvre. ILO evidence indicates that casualization of agricultural wage labour is a worldwide tendency. The WDR 2008 admits that agricultural wage labourers have been known to face highly exploitative working conditions. Nonetheless, whereas the WDR 2008 calls for farmers to form producers' organizations to pursue their collective interests, there is no such call for agricultural labourers' trade unions to fight for decent wages and security of employment.

Labour uncertainties of smallholder household members

One of the paradoxes of rural dwellers' non-agricultural income diversification is that it tends to be triggered in part by capital constraints in commercial agriculture, while success in non-agricultural activities is largely determined by access to capital. The challenge is to source sufficient starting capital and avoid the ever-present danger of running down one's working capital by using it for necessary consumption rather than business operations. Complicated multiple livelihood strategies arise from the fact that neither subsistence production, agricultural commodity production nor non-agricultural activities in and of themselves provide security of livelihood. The search for the right balance of activities becomes incessant.

Through work experimentation and income diversification rural households have become inexorably entwined in a process of economic differentiation. Easy-entry activities requiring relatively little capital investment, such as women's beer-brewing, are quickly over-saturated, triggering a high rate of economic failure. Those in the higher capital entry activities or with better physical mobility, which affords them access to other markets, are far more likely to succeed. In this way, income diversification staves off the hunger that would have undoubtedly become more problematic in rural areas.

The neo-liberal policy paradox has been one in which the fluidity of commodities and labour has been equated with economic opportunity and prosperity, whereas the reality for vast numbers of people has been economic decline. Little attention has been accorded to the political and social stability of African rural societies in relation to the pervasive threat of labour redundancy and growing impoverishment. In some countries and regions, increased labour mobility has harboured ethnic tension. Ethnic ties, useful for economic networking, have within them an 'us versus them' tendency. Some groups, such as certain groups of rural traders or miners, are likely to succeed, and may even be in a position to flaunt their wealth, whereas the majority who remain as family farmers will experience economic marginalization as 'relic farmers'. These economic tensions should not be delinked from the political sensitivities associated with the future for African smallholders and their nation-states in the global market.

Reading Between the Lines of the WDR 2008: African Smallholders' Rural Future

Abandoning food security and smallholder agricultural development aims

Comparing the content of the WDR 1982 and the WDR 2008, both share the theme of agricultural development, but depart radically from each other in their analysis of African smallholder agricultural prospects and policy recommendations. International efforts to eliminate world hunger were the backdrop to the 1982 document. Regardless of the oil crisis of the 1970s and the world economic upheaval in its aftermath, the WDR 1982 was upbeat. International research stations' efforts to boost staple food yields which began in the 1960s, coalesced into the green revolution with profound leaps in agricultural productivity and consequent welfare-enhancing effects on the Asian countryside during the 1970s.[1] The WDR 1982 not only endorsed but also assumed that similar developments could take place in Sub-Saharan Africa following further investment, research and extension to achieve the goal of ridding Africa of food insecurity.

---

1. The Rockefeller Foundation funded a team to develop improved maize varieties as early as in 1943, through crop research stations. In 1960, the International Rice Research Institute (IRRI) was established in the Philippines supported by the Rockefeller and Ford Foundations and the Centro International de Majoramiento de Maiz y Trigo (CIMMYT) followed for maize in Mexico in 1966. A string of other research stations followed thereafter including the International Institute of Tropical Agriculture (IITA), the International Crops Research Institute for the Semi-Arid Tropics (ICRISAT) in Nigeria and the International Centre for Agricultural Research in the Dry Areas (ICARDA). The World Bank has played a key role, providing the chairmanship and secretariat for the Consultative Group on International Research (CGIAR) formed in 1971 as an association linking these various research institutions (WDR 1982:68).

Research and technology need to be developed and adapted to local conditions. The lack of technological improvements suitable for African conditions is a main reason for Africa's poor performance so far. (WDR 1982: 91).

There was a time lag of roughly ten years between the initiation of research and extension for the green revolution in Asia and the embryonic green revolution efforts that took place in Africa. However, there were signs that even before the WDR 1982 was published, the World Bank was already pulling support away from green revolution efforts in Africa. The World Bank's 1981 Berg report, which specifically addressed the African continent's deepening debt crisis, proposed an entirely different policy from that of green revolution-directed agricultural investment. Concern for African food security was jettisoned. The Berg report was the blueprint for the structural adjustment programmes whose cutbacks aimed at tightening belts rather than filling bellies with food. The yield improvements that had so far been achieved on the basis of improved maize varieties in East and Southern Africa documented by Eicher (1995 and 2001) could not progress. Agronomic research and extension cutbacks dissipated the momentum towards resolving food insecurity. Thus, the agrarian solution to African hunger did not receive international research effort on a par with that accorded to Asia. African agriculture entered a protracted period of stagnation. Agricultural funding slipped from being 24 per cent of total World Bank lending[2] in 1982 to 12 per cent in 1997–99 – a funding trend mirrored by many other western bilateral donors and African states.

Twenty-six years later, the WDR 2008 is a far less analytical document than that of 1982. The WDR 2008 suffers from a logical inconsistency between its acclaimed goal of poverty alleviation for African smallholder farmers and its conviction that large-scale commercial farming is the inevitable future of farming. African small-scale family farmers must meet the productivity levels, rigorous product standards and delivery schedules of international value chains to compete effectively, yet without necessary support. The following quote from the draft WDR 2008 maps the road ahead for smallholders.

> An emerging vision of agriculture for development redefines the roles of producers, the private sector, and the state. Production is mainly by smallholders, who often remain the most efficient producers, in particular when supported

---

2. Having been as high as 32 per cent in 1976–78 (Pincus 2001).

by their organizations. But when these organizations cannot capture economies of scale in production and marketing, labor-intensive commercial farming can be a better form of production, and efficient and fair labor markets are the key instruments to reducing rural poverty. The private sector drives the organization of value chains that bring the market to smallholders and commercial farms. The state – through enhanced capacity and new forms of governance – corrects market failures, regulates competition, and engages strategically in public-private partnerships to promote competitiveness in the agribusiness sector and support the greater inclusion of smallholders and rural workers. In this emerging vision, agriculture assumes a prominent role in the development agenda (WDR 2008, Overview, ch.1, page 11, July version).

At present hundreds of millions of African peasant smallholders are not competing successfully in global commodity markets. The World Bank adopts a matter-of-fact position that they will relinquish their autonomy as agricultural producers and work as contract farmers or wage labourers in large-scale agribusiness or alternatively leave agriculture to seek their livelihood elsewhere. Their sanguine attitude towards peasant labour redundancy does not tally with their professed concern for the African rural poor. Beneath the WDR 2008's public relations spin about poverty alleviation, they are conferring *carte blanche* support to a 'survival of the fittest' economic trajectory in which the grossly imbalanced commercial interests of large-scale OECD subsidized farmers, supermarket chains and agribusiness have full scope to compete against unsubsidized peasant farmers engaged in rural ways of life that have managed hitherto to endure for millennia.

As has been argued in preceding sections, the World Bank has not been held accountable for the agricultural policy misjudgements and blunders they have enforced in Africa over the last 25 years through structural adjustment policy and debt conditionality. Now, with impunity, they are throwing their weight behind the rapid massive redundancy of peasant smallholders in the name of African development. In the process, they irresponsibly overlook the likely economic, political and social consequences of policies that assault the cultural and economic bedrock of African nation-states – their agrarian roots. This is market fundamentalism on the rampage.

It is useful to draw back and consider why the World Bank has not opted for supporting the re-kindling of agricultural research and input subsidies for African peasants along the lines of the green revolution of Asia, even though they readily admit that this was foundational to Asia's current economic success. There are various clues in the text of the WDR 2008.

First of all, it is inferred that there is a lack of political will to generate the international funding. The Cold War political impetus is gone. Western fears that Asian nations, which were subject to the pressures of their massive hungry rural populations, could succumb to Soviet influence during the 1960s are now history.

Second, in the context of the biotechnology revolution, agronomic research is now increasingly privatized. Promising advances are quickly patented to ensure future profits and state-funded agronomic research does not have the same capability as it did forty years ago. Finally, and most significantly, the World Bank, in its current phase of extreme market fundamentalism, has an aversion to state intervention for agricultural input and service delivery, even though it acknowledges the success of such a delivery mode during the Asian green revolution. The African state is not to be entrusted with providing the direction and infrastructural and service support that would be required over ten to fifteen years to ensure the success of a green revolution.

But is the potential for a green revolution in Africa now so different than that of Asia in the1960s? The need is just as pressing. Africa's rural and urban populations are expanding. The FAO reports that 200 million Africans go to bed each night hungry . They do so amidst what is termed an international 'war on terror'. North and East Africa have already been identified as hot spots for recruitment as well as sites of terrorist attacks. Any part of the African continent experiencing deepening poverty and hunger could be especially vulnerable to terrorist activity. How different is this from the Cold War threat in the 1960s?

Second, while private interests in agronomic research may have intensified since the 1960s, there is, just as before, a great deal of public humanitarian support, emanating from private corporate finance, is not only interested but eager to contribute to a 'cure' for world hunger. The Sasakawa Foundation and more recently the Bill and Melinda Gates Foundation are now investing heavily and beginning to register success. Furthermore, private corporate interests in agronomic research seeking to safeguard their findings for profit maximization, may nonetheless be open to some 'sharing', as illustrated by the recent example of international drug firms dramatically reducing their prices for anti-retrovirals (ARVs) for AIDS treatment in African countries under international humanitarian pressure and competition from companies selling generic versions of the ARVs.

Finally, and most importantly, while the World Bank continues to perceive African states as inefficient and corrupt and thereby incapable of service and infrastructure delivery for a green revolution, it is overlooked that international concerns about levels of corruption in Asian states were just as rife on the eve of the green revolution there (Olukoshi 1998, Myrdal, 1968).

Some form of an agricultural revolution is vital to the future of today's African smallholders (Djurfeldt et al., 2005). This is not because of their need to remain in the agricultural sector, although this may be the desire of many. Rather it is because the food security afforded by a green revolution provides the necessary foundation and insurance for individuals, rural households and nation-states to develop non-agrarian occupational specializations as well as constituting an important impetus for the growth of other sectors (Bhaduri and Skarstein, 1997).

Sequenced timing of depeasantization

SAPs and economic liberalization have set in train a process of accelerated depeasantization which is causing peasant households to disintegrate as coherent economic, social and cultural units. This process has been unfolding over the last 25 years. A generation of rural youth have spent their entire lives in this transitional process, no longer bounded so firmly by local traditional authorities and adherence to strict intra-household gender/age divisions of labour (Thorsen, 2007). They have been the most active participants in the scramble for alternative livelihoods as reported earlier. They are familiar with the many constraints of working in the informal sector on an exploratory basis and have witnessed growing economic differentiation. They face continual crippling limitations of capital to start up and maintain their livelihood activities.

This youthful transitional generation, 'the scrambled generation', have in many contexts been highly mobile, often moving between the countryside and city. Given their locational and occupational 'otherness' vis-à-vis their rural home areas they constitute a link between rural and urban areas, and between agrarian and non-agrarian pursuits. For the most part, they have coped with the challenges remarkably well and have built up a skill base, particularly in trade, that has not hitherto been so prevalent.

Nonetheless, in some places, the strains have been too great and the insecurities of seeking a livelihood combined with national political tensions have broken out in civil wars. There have been many examples over the last 25 years: Uganda, Rwanda, Somalia, Chad, the Democratic Republic of Congo, Sudan, and Sierra Leone, to name but a few (Buijtenhuis, 2000). In these wars, young men may find lucrative livelihoods that supercede the earnings they could otherwise eke out. What this points to is that the experiences of depeasantization at an individual and household level are entwined with the national level in both positive and negative ways. Individuals are disengaging from their rural household context, and all the certainties of locational residence, domestic security and farming as a lifetime occupation. Individual identities may become disoriented, particularly if the rate of change and the level of insecurity are high. When the individual fails to find a viable livelihood with decent returns and the dignity of work, demoralization may begin to set in. Men in particular are liable to react to livelihood disorientation and seek escape through alcohol, drugs, crime or suicide.

Depeasantization is already underway in rural Africa and, however difficult it is for individuals and rural households in the short and medium term, it can deliver positive long-term outcomes in terms of higher paid labour pursuits and new fulfilling ways of life. But this is only likely when the pace of change is moderate and the working conditions and remuneration in the process are not overly demeaning. It is in this light that we can turn to the World Bank's recommended pathways for African smallholders.

Well-charted predictable developmental path or descent into deepening poverty?

Depeasantization has happened time and again in world history over the last three hundred years in one country after another. African depeasantization however is different from most parts of the world which have undergone the process because it has not experienced an agricultural revolution to lift its agricultural productivity nor is it undergoing an industrial revolution to raise its non-agrarian productivity. The latter is unlikely to occur for the foreseeable future. Thus the World Bank is recommending global capital's destruction of an independent smallholder agricultural sector in the absence of clear employment prospects. This is radically different from the rapid depeasantization process currently underway in China. There, members of rural households are leaving the farm to work in the booming industrial and service sectors of the national economy. Given the constricted parameters of African national economies, smallholder alternative options outlined by the World Bank are not convincing.

African smallholder export and food production

In smallholder export agriculture, there is little scope for unsubsidized African farmers to compete. The WDR 2008's proposed limited support for 'smart' fertilizer subsidies pales in the face of the large producer subsidies that the heavily capitalized farmers in most OECD countries receive. The African farmers cannot compete because they are not on a level playing field. As a consequence they are, in effect, out of the game. They are not even on the bench waiting to play anymore. Figure 1 showed them at the bottom of the heap in terms of income-earning. They are now dropping off the graph.

The WDR 2008 stresses that there are market opportunities for farmers to feed the cities. This is a much stronger possibility than export crops, but as an earlier section of this analysis has outlined, liberalized staple food markets are fraught with pitfalls for the smallholder producer whereas the large-scale domestic producers are likely to be less vulnerable to the staple food market fluctuations. Finally in unprotected national food markets both are subject to the possibility of stiff competition from subsidized grain producers from abroad, particularly if the concentrations of demand, notably the main cities, are located on the coast, where cheap food imports can be shipped at a fraction of the transport costs of bringing supplies from remote up-country areas of the country.

Alternatives to smallholder production

Contract farming and agricultural wage labour have been proposed as alternatives to own-farm production as smallholders. This is outlined as a ready alternative which, when accompanied with fair remuneration and working conditions, is a productive and just solution. The question remains how such just conditions are to be secured. Large-scale farms and agri-business are not charities. A deluge of farmers, exiting the smallholder sector as 'refugees', and flooding rural labour markets, will meet with extremely low returns and harsh working conditions. The WDR 2008 avoids mentioning the role of labour trade unions, which is a glaring oversight.

Rural non-agricultural activities are performed primarily on the basis of self-employment. The risks are high and financial capital and over-supply are the over-riding constraints. The rural informal sector is already heavily over-subscribed and known for its low, unreliable fluctuating levels of remuneration. Finally, there is the option to migrate to an urban area to seek employment. In most cases the outcome is very similar to that of participating in rural non-farm activities without the safety net of having farming members of the family nearby.

Provision for those without an alternative: African
'rural holding grounds'

The WDR 2008 advocates the above listed options as escape routes to avoid directly experiencing the disintegration of peasant smallholder farming, but there is a realization that not all African rural dwellers will manage to join the exodus. For those who are left behind, the policy will be 'social protection' rather than 'economic development'. In this sense the WDR 2008 marks a major departure in World Bank rural policy – African rural development policy will no longer centre on smallholder agency. Rather those who constitute the 'relic population', will be availed, a continued subsistence[3] farming base, facilitated by the World Bank's recent switch to acceptance of the historical evolution of customary tribal-based land tenure. In other words, those left in the countryside will live on tribal communal 'holding grounds', akin to the Bantustans of the apartheid period of South African history, eking out an existence on the basis of exceptionally low-yielding, uncapitalized agriculture . Like the Bantustans, these holding grounds could function as labour reserves for the mainstream national economy and would most likely be based on conservative tribal customary legal frameworks not only with respect to land but in a wide array of other spheres as well. It is indeed an irony that such a model resurfaces little more

---

3. As opposed to peasant farming which definitionally entails the combination of subsistence and commodity agricultural production.

than a decade after South Africa managed to rid itself of this 'separate and unequal' model of rural exploitation in the name of development.

Another analogy surfaces. Nineteenth century North American Indians, practising a hunting and gathering mode of livelihood were deemed backward and in the way of the land requirements for incoming migrant European settler farming. Decades of Indian wars culminated in tribes across North America giving up their time-honoured mode of livelihood and being corralled into 'reservations' where the collective demoralization arising from the loss of their economic livelihood and cultural identity can be witnessed to this day by the high rates of alcoholism, poor health and suicide, especially amongst the male population of the reservation.

In essence, the juncture that rural Africa now faces is that of continuing on the path of depeasantization at the accelerated rate dictated by a politically imbalanced global market ready for the kill, or allowing African states to depeasantize at a slower rate and in a far more constructive manner.

Conclusion

Over the past three decades African smallholder farming has contended with serious erosion from an inhospitable global market and unsupportive state policies chipping away at its productive infrastructure, services and incentives. However, African smallholder farming continues to have land, labour and local institutions at its disposal. It needs timely support and capital injections in the form of investment in research, extension, infrastructure, improved inputs and enhancement of mutual learning processes amongst farmers. Agricultural productivity and marketing improvement objectives would be similar to those achieved by the Asian green revolution but different in terms of promoting African smallholder knowledge and productive capacities in an environmentally friendly way. Such developments are not new to Africa. They were already emerging in the 1970s (Eicher, 1995 and 2001; Eicher and Kupfuma, 1997; Djurfeldt et al. 2005) but were shortcircuited by economic crisis and structural adjustment. Now, however, a renewed momentum is underway as evidenced by the efforts of amongst many others, Sasakawa 2000 and, most recently, the Alliance for a Green Revolution in Africa [AGRA].[1]

To be effective, however, the approach to African agricultural development has to be based on a thorough understanding of local smallholder rural institutional settings, including the gender and inter-generational relationships, and rural-urban interconnections. This implies that the social, cultural and political dimensions of agrarian change, including state-smallholder relationships, cannot be ignored. Further, efforts have to be open to timely measures to subsidize and protect smallholder farmers and their organisations to give them the economic means, motivation and self-esteem

---

1. The Bill and Melinda Gates Foundation funded agricultural research programme. See http://agra-alliance.org. For criticism of a biotechnology-based 'green revolution' in Africa see: www.grain.org and www.etcgroup.org.

to produce for national staple food markets and to compete more fairly with capitalized farmers elsewhere. These measures have to be individually tailored to the many agricultural and food production systems of the continent.

Much of the controversy surrounding the concept of an African green revolution centres on the role of the African state as a programme executor and service delivery agent and the degree of attention to environmentally sustainable agricultural practices. These are issues that have to be resolved with respect to the historical background and geographical specificities of the country and rural locality in question. The search for the 'correct' approach will only serve to delay action weakening African smallholder farming still further relative to large-scale agricultural production.

Highlighting the perilous predicament of smallholders, the World Bank is finally taking cognizance of the growing economic and political risks facing smallholders and their governments. The World Bank has recently initiated a study of the implications of liberalization and global value chains on rural development in developing countries.[2] However it remains to be seen if the findings have any influence on the World Bank's market fundamentalist policies.

In the meantime, current efforts aimed to improve conventional plant breeding techniques, have the advantages that they can deliver significant benefits in the short run and fit well with the regulatory frameworks in place in most African countries. They create scope for pooling the knowledge of scientists and smallholders, giving smallholders a central role in the promotion of African agriculture and continue to exercise influence and control over its development (AGRA, 2007 and Busch, 1997). This contrasts with the WDR 2008's focus on accommodating large-scale commercial farming and vertical agricultural value chains structured by agribusiness and supermarkets.

Considerable investment is required to reinvigorate smallholder African agriculture. This is critical not only to smallholder welfare but to national economic development – providing the necessary foundation for occupational self-esteem and work identities and political stability and a sense of basic security upon which a strong non-agrarian future can be built.

---

2. This is a two-year (2006–2008) cross-regional research programme within the Sustainable Development Department of the World Bank funded by the French Ministry of Foreign and European Affairs and IFAD.

Conclusion

Unless this comes about, African agriculture and rural areas will constitute a vast 'holding ground' of immense social and economic misery with potential dramatic impacts on global politics, migration and environmental and climatic aspects.

References

Adams, M., S. Sibanda and S. Turner, 1999, "Land Tenure Reform and Rural Livelihoods in Southern Africa", *ODI Natural Resource Perspectives*, Number 39.

African Biodiversity Network, 2007, "Agrofuels in Africa – the Impacts on Land, Food and Forests". Case studies from Benin, Tanzania, Uganda and Zambia. July.

Alliance for a Green Revolution in Africa (AGRA), 2007, "Plant Breeding and Genetic Engineering", http://www.agra-alliance.org/about/genetic_engineering.html.

Altieri, M. and E. Bravo, 2007, "The ecological and social tragedy of crop-based biofuel production in the Americas", April 2007. http://foodfirst.org/node/1662.

Asian Development Bank (ADB), 2002, "The Value Chain for Tea in Vietnam: Prospects for Participation by the Poor". ADB Discussion Paper No. 1.

Atakilte Beyene, 2003, *Soil Conservation, Land Use and Property Rights in Northern Ethiopia. Understanding environmental change in smallholder farming systems*. Phd Dissertation. Agraria 395, Swedish University of Agricultural Sciences, Uppsala, Sweden.

Berry, S., 1993, *No Condition Is Perma-nent: The Social Dynamics of Agrarian Change in Sub-Saharan Africa*. The University of Wisconsin Press, East Lansing, Michigan.

Bhaduri, A., 1986, "Forced Commerce and Agrarian Growth", *World Development*, Vol 14, No 2, pp. 267–272.

Bhaduri, A. and R. Skarstein, 1997, *Economic Development and Agricultural Productivity*. Edward Elgar Publishing Ltd., Cheltenham, UK.

Bilame, O.S., 1996, *Performance of maize production during structural adjustment programmes in Tanzania*. MA thesis, University of Dar es Salaam.

Bratton, M. and N. van de Walle, 1997, *Democatic Experiments in Africa – Regime Transitions in Comparative Perspective*. Cambridge University Press, Cambridge.

Bruce, J.W. and S.E. Mighot-Adholla (eds), 1994, *Searching for Land Tenure Security in Africa*. Kendall/Hunt, Dubuque, Iowa.

Bryceson, D.F., 2000, "Peasant Theories and Smallholder Policies: Past and Present", Bryceson, D.F., C. Kay and J. Mooij (eds), *Disappearing Peasantries? Rural Labour in Africa, Asia and Latin America*. London, ITDG Publishing, pp. 1–36.

—, 2002a, "The Scramble in Africa: Reorienting Rural Livelihoods", *World Development* 30 (5), 725–39.

—, 2002b, "Multiplex Livelihoods in Rural Africa: Recasting the Terms and Conditions of Gainful Employment", *Journal of Modern African Studies* 40 (1), 1–28.

References

—, 2006, "Vulnerability and Viability of East and Southern Africa's Apex Cities", Bryceson, D.F. and D. Potts (eds), *African Urban Economies*. New York, Palgrave Macmillan, pp. 319–40.

Bryceson, D.F. and L. Bank, 2001, "End of an Era: Africa's Development Policy Parallax", *Journal of Contemporary African Studies*, 19 (1), pp. 5–23.

Buijtenhuis, R., 2000, "Peasant Wars in Africa: Gone with the Wind?" Bryceson, D.F., C. Kay and J. Mooij (eds), *Disappearing Peasantries? Rural Labour in Africa, Asia and Latin America*. London, IT Pulications, pp. 112–122.

Busch, L., 1997, "Biotechnology and agricultural productivity: Changing the rules of the game?" in Bhaduri, A. and R. Skarstein (eds), op. cit., pp. 242–255.

Chabal, P. and J-P. Daloz, 1999, *Africa Works – Disorder as Political Instrument*. The International African Institute with James Currey and Indiana University Press, Oxford and Bloomington.

Chanock, M., 1985, *Law, Custom and Social Order*. Cambridge University Press, Cambridge.

Chenery, H., M.S. Ahluwalia, C.L.G. Bell, J.H. Duloy and R. Jolly, 1974, *Redistribution with Growth*. Published for the World Bank and the Institute of Development Studies, University of Sussex by Oxford University Press.

Childs, N., 2005, *Rice Situation and Outlook Yearbook*. US Deptartment of Agriculture, Economic Research Service, RCS-2005.

Clapham, C., 1985, *Third World Politics – An Introduction*. University of Wisconsin Press, Madison.

Comar, V. and J.M. Gusman Ferraz, 2007, "Brazil's Sugar Cane Ethanol: Villain or Panacea". Institute for Environment and Development (IMAD) and EMBRAPA/CNPMA, mimeo.

Cousins, B. and A. Claassens, 2006, "More than simply 'socially embedded': Recognizing the distinctiveness of African land rights", keynote address at the international symposium on "At the frontier of land issues: Social embeddedness of rights and public policy". Montpellier, May 17–19, Paris, France.

DAC, 1983, "Increasing the effectiveness of development co-operation through improved co-ordination at the country level". Note by the Secretariat, OECD, Paris.

Delgado, C.L., N. Minot and C. Courbois, 1999, *Agriculture in Tanzania since 1986: Follower or Leader of Growth?* Revised version, International Food Research Institute, Washington DC, November.

Djurfeldt, G., H. Holmén, M. Jirstrom, and R. Larsson, 2005, *The African Food Crisis: Lessons from the Asian Green Revolution*. CABI Publishing, UK.

Duveskog, D., 2007, "New thinking in Agricultural Extension – with focus on East Africa. A literature review". Department of Urban and Rural Development, the Swedish University of Agrcultural Sciences, Uppsala.

Easterly, W., 2003, "Can Foreign Aid Buy Growth?", *Journal of Economic Perspectives*, Vol 17, No 3, pp. 23–48.

Eicher, C. K., 1995, "Zimbabwe's maize-based Green Revolution: Preconditions for replication", *World Development*, 23, 805–18.

—, 2001, "Africa's unfinished business: Building sustainable agricultural research systems". Staff paper no. 2001–10. Department of Agricultural Economics, Michigan State University, East Lansing, Michigan.

Eicher, C.K. and B. Kupfuma, 1997, "Zimbabwe's emerging maize revolution", in Byerlee, D. and C.K. Eicher (eds), *Africa's Emerging Maize Revolution*. Lynne Rienner Publishers, Boulder, Colorado, pp. 25–43.

Elbadawi, I. and A. Gelb, 2003, "Financing Africa's Development – Towards a Business Plan?", in van de Walle, N., N. Ball and V. Raamachandran, *Beyond Structural Adjustment*. Palgrave/Macmillan, New York and Houndsmill, pp. 35–75.

Ellis, F. and H. A. Freeman, 2005, "Comparative Evidence from Four African Countries", Ellis, F. and H. A. Freeman (eds), *Rural Livelihoods and Poverty Reduction Policies*. London, Routledge, pp. 31-47.

FAO, 2007, "Growing bio-fuel demand underpinning higher agriculture prices". FAO Newsroom, July 4.

Fernando-Cornejo, J. and M. Caswell, 2006, *The First Decade of Genetically Modified Crops in the United States*. US Department of Agriculture, Economic Research Service, Economic Information Bulletin 11.

Fortucci, P., 2003, *The Contributions of Cotton to Economy and Food Security in Developing Countries*. United Nations Food and Agriculture Organisation (FAO), June.

Fundanga, C.M., 1996, "Practical Effects of Economic and Political Conditionality in the Recipient Administration". K. Havnevik and B. van Arkadie (eds), op. cit., pp. 89–97.

Gabre-Madhin, E., 2006, "Building Institutions for Markets: The Challenge in the Age of Globalization", mimeo, EGDI, Swedish Ministry for Foreign Affairs, Stockholm.

Gabre-Madhin, E., C.B. Barret and P. Dorosh, 2003, *Technological Change and Price Effects in Agriculture: Coneceptual and Comparative Perspectives*. MTID Discussion Paper no. 62, International Food Policy Research Institute, Washington D.C. April 2003.

Gibbon, P., 1992, "The World Bank and African Poverty 1973–91", *Journal of Modern African Studies*, Vol 30, No 2.

—, 2006, *Current Trends and the New Development Role of Commodities*. Amsterdam, Common Fund for Commodities.

Gibbon, P., K. Havnevik and K. Hermele, 1993, *A Blighted Harvest. The World Bank and African Agriculture in the 1980s*. London and Trenton NJ, James Currey and Africa World Press.

Gibbon, P. and S. Ponte, 2005, *Trading Down: Africa, Value Chains and the Global Economy*. Philadelphia, Temple University Press.

Gudoshnikov, S., L. Jolly and D. Spence, 2004, *The International Sugar Market*, Woodhead Publishing, Cambridge.

Gunning, J.W., 2004, "Why Give Aid", Free University, Amsterdam, paper prepared for the 2nd AFD-EUDN Conference, Paris, Nov. 25.

Havnevik, K.J. (ed.), 1987, *The IMF and the World Bank in Africa. Conditionality, Impact and Alternatives*. Seminar Proceedings No 18. Scandinavian Institute of African Studies (The Nordic Africa Institute), Uppsala.

Havnevik, K. and B. van Arkadie (eds), 1996, *Domination or Dialogue. Experiences and Prospects for African Development Cooperation*. The Nordic Africa Institute, Uppsala.

Havnevik, K.J., 2000, "The Institutional Heart or Rural Africa", in K.J. Havnevik with E. Sandström (eds), *The Institutional Context of Poverty Eradication in Rural Africa*. Proceedings from a seminar in tribute to the 20th Anniversary of the International Fund for Agricultural Development, IFAD, The Nordic Africa Institute, Uppsala, pp. 38–50.

Havnevik, K., T. Negash and A. Beyene (eds), 2006, *Of Global Concern – Rural Livelihood Dynamics and Natura Resource Governance*, Sida Studies no. 16. Sida, Stockholm.

Hawassi, F.G.H., N.S.Y. Mdoe and F.M. Turuka, 1999, "Efficiency in fertilizer use among smallholder farmers in Mbinga district", *AGREST Proceedings*, Sokoine University of Agriculture, pp. 72–86.

IFAD, 2002, "The Rural Poor. Survival or a better life?". Paper submitted by IFAD to the World Summit on Sustainable Development, South Africa, August 26 to September 4.

Izumi, K., 2006, *Land and Property Rights of Women and Orphans in the context of HIV and AIDS. The Case Study of Zimbabwe*. Cape Town: HSRC Press.

—, 2007, *Reclaiming Our Lives: HIV and AIDS, Women's Land and Property Rights and Livelihoods in Southern and East Africa – Narratives and Responses*. Cape Town, HSRC Press.

Jua, N. and P. Nkwisi (eds), 2001, "Levelling the Playing Field: Combating Racism, Ethnicity and Different forms of Discrimination". MOSTEHNO-NET African Publications, UNESCO/END.

Kairaba A., 2002, "Rwanda Country Case Study". Paper presented at the Regional Workshop on Land Issues in Africa and the Middle East, Kampala, Uganda, 29 April–2 May.

Kox, H., 2000, "The Market for Cocoa Powder". Report prepared for the Netherlands Ministry of Foreign Affairs.

Lipset, S.M., 1984, "Social Conflict, Legitimacy and Democracy", in Connolly, W. (ed.), *Legitimacy and the State*. Basil Blackwell, Oxford.

MacDonald, J. et al., 2004, "Contracts, Markets and Prices: Organising the Production and Use of Agricultural Commodities". US Department of Agriculture, Economic Report No. 837.

Mamdani, M., 1996, *Citizen and Subject – Contemporary Africa and the Legacy of Late Colonialism*. Princeton University Press, Princeton.

Manji, A., 2002, *The World Bank's Policy Research Report 'Land Policy for Pro Poor Development': A Gender Analysis*. (www.oxfam.org.uk/landrigths/manjiPRR.doc).

Matondi, P.B., 2001, *The Struggle for Access to Land and Water Resources in Zimbabwe: The Case of Shamva District*. Doctoral Thesis, Department of Rural Development Studies, Swedish University of Agricultural Sciences, Uppsala, Sweden.

Myrdal, G., 1982, *Asian drama: An inquiry into the poverty of nations*. New Delhi, Kalyani Publishers.

Norfolk, S., and D. Soberano, June 2000, "From conflict to partnership. Relationships and land in Zambézia Province", Land Tenure Component, ZADP, Quelimane.

North, D.C., 1993, *Institutionerna, tillväxten och välståndet*. SNS förlag, Stockholm.

North, D.C and R.P. Thomas, 1970, "An Economic Theory of the Growth of the Western World", *The Economic History Review*, Vol 23, No 1, pp 1–17.

—, 1971, "The Rise and Fall of the Manorial System: A Theoretical Model, *Journal of Economic History*, Vol 31, pp. 777–803.

Ntsebeza, L., 2004, "Democratic Decentralisation and Traditional Authority: Dilemmas of Land Administration in Rural South Africa", *European Journal of Development Research*, Vol 16, No 1, pp. 71–89.

Olukoshi, A., 1998, *The Elusive Prince of Denmark. Structural Adjustment and the Crisis of Governance in Africa*. Research Report No. 104, The Nordic Africa Institute, Uppsala.

Oxfam, 2002, *The Tea Market: A Background Study. Draft for Comments*.

Øygard, R. et al., 2003, *The maze of maize: Improving input and output market access for poor smallholders in the Southern Africa region – The experience of Zambia and Malawi*. Typescript, Noragric, Ås, June.

Peck, T., 2001, *The International Timber Trade*. Woodhead Publishing, Cambridge.

Peters, P., 2002, "The Limits of Negotiability: Security, Equity and Class Formation in Africa's Land Systems", in Juul K. and C. Lund (eds), *Negotiating Property in Africa*. Portsmouth, NH, Heinemann.

Pincus, J., 2001, "The Post-Washington Consensus and Lending Operations in Agriculture: New Rhetoric and Old Operational Realities", in Fine, B., C. Lapavitsas and J. Pincus (eds), *Development Policy in the Twenty-First Century: Beyond the Post-Washington Consensus*. London, Routledge.

Platteau, J-P., 1996, "The Evolutionary Theory of Land Rights as Applied to Sub-Saharan Africa: A Critical Assessment", *Development and Change*, Vol 27, No 1, pp. 29–86.

Reardon, T., 1997, "Using Evidence of Household Income Diversification to Inform Study of the Rural Nonfarm Labor Market in Africa", *World Development* 25 (5), pp. 735–48.

Mkandawire, T., 1996, "Economic Policy-Making and the Consolidation of Democratic Institutions in Africa", in Havnevik, K. and B. van Arkadie (eds), *Domination or Dialogue. Experiences and Prospects for African Development Cooperation*. The Nordic Africa Institute, Uppsala, pp. 24–48.

Rothstein, B., 1994, *Vad bör staten göra? – Om välfärdstatens moraliska och politiska logik*. SNS förlag, Stockholm.

Rukuni, M., P. Tawonezvi, C. Eicher, with M. Munyuki-Hungwe and P. Matondi (eds), 2006, *Zimbabwe's Agricultural Revolution*. Second edition, University of Zimbabwe Publications.

Sagasti, F., K. Bezanson and F. Prada, 2005, *The Future of Development Financing, Challenges and Strategic Choices*. Palgrave/Macmillan, Houndmills.

Selby, A., 2006, *Commercial farmers and the state: Interest group politics and land reform in Zimbabwe*. PhD thesis, Oxford University, United Kingdom.

Skarstein, R., 2005, "The Tanzanian smallholder under the yoke of liberalisation. From bad to worse?" Paper presented at international conference, *Tanzania – A Critical Analysis*, The Nordic Africa Institute, Uppsala.

Thorsen, D., 2007, "Junior-Senior Linkages. Youngsters' Perceptions of Migration in Rural Burkina Faso", in Hanh, H.P. and G. Klute (eds), *Cultures of Migration – African Perspectives*. Munster, Lit Verlag, chapter 9.

Torkelsson, Å., forthcoming, "Resources, Not Capital. A Case Study on the Gendered Distribution and Productivity of Social Resources in Rural Ethiopia", *Rural Sociology*.

UNCTAD, 2006, *Trade and Development Report: Global Partnerships and National Policies for Development*, Geneva.

UNDP/World Bank, 1989, *Africa's Adjustment and Growth in the 1980s*. Washington DC.

UNICEF, 1987, *Adjustment with a Human Face*. New York.

Unnevehr, L. and N. Hirschhorn, 2000, *Food Safety Issues in the Developing World*. World Bank Technical Paper 469.

URT/MAC (United Republic of Tanzania – Ministry of Agriculture), 2000, *Basic Data – Agriculture and Livestock Sector 1992/93 – 1998/99*. MAC, Dar es Salaam.

van Dijk, J., D. van Doesburg, A. Heijbroek, M. Wazir, and G. de Wolff, 1998, *The World Coffee Market*. Rabobank, Utrecht.

Van Donge, K.J., 1999, "Law and order as a development issue: Land conflicts and the creation of social order in Malawi", *Journal of Development Studies*, Vol 36, No 2.

Wade, R., 2001, "Showdown at the World Bank", *New Left Review*, 7, January/February.

Wangwe, S., 1987, "Impact of the IMF/World Bank Philosophy, the Case of Tanzania", in Havnevik, K. (ed.), 1987, op. cit.

World Bank, 1981, *Accelerated Development in Sub-Saharan Africa: An Agenda for Action*, Washington DC, World Bank.

—, 1982, *World Development Report 1982 – International Development Trends and Agriculture and Economic Development*. Washington DC, World Bank.

—, 1989, *Sub-Saharan Africa: From Crisis to Sustainable Growth*. Washington DC, World Bank.

—, 1995, *Workers in an integrating world: World development indicators*. New York, Oxford University Press.

—, 2000a, *Voices of the Poor – Crying out for Change*. Washington DC, World Bank.

—, 2000b, *Agriculture in Tanzania since 1986 – Follower or leader in growth?* World Bank Report No. 20639, Washington DC.

—, 2000c, *Can Africa Claim the 21st Century?* Washington DC, World Bank.

—, 2001, *World Development Report 2000/2001. Attacking Poverty.* Published for the World Bank by Oxford University Press.

—, 2003, *Land policies for growth and poverty reduction*. Washington DC, World Bank.

—, 2007, *World Development Indicators*, https://publications.worldbank.org/subscriptions/WDI/old-default.htm.

—, 2008, *World Development Report – Agriculture for Development*. Washington DC, World Bank.

WTO, 2004, Document WT/L/579, 2 August.

Zagah, R., 2005, *Economic Growth in the 1990s – Lessons from a Decade of Reform*. Washington DC, World Bank.

Acronyms

ARV	Anti-retroviral
ASAR	African Sustainable Agricultural Revolution
CGIAR	Consultative Group on International Agricultural Research
CIMMYT	Centro International de Maiz y Trigo
DSB	Dispute Settlement Body
FAO	Food and Agricultural Organisation of the UN
GDP	Gross Domestic Product
HIPC	Highly Indebted Poor Countries
ICRISAT	the international Centre for Agricultural Research in the Dry Areas
IFAD	International Fund for Agricultural Development
IITA	the International Institute of Tropical Agriculture
ILO	International Labour Organisation of the UN
IMF	International Monetary Fund
IRRI	International Rice Research Institute
NGO	Non-governmental Organisation
NIEO	New International Economic Order
OECD	Organisation for Economic Cooperation and Development
SAP	Structural Adjustment Programme
UNCTAD	United Nations Conference on Trade and Development
UNDP	United Nations Development Programme
UNICEF	The United Nations Children's Fund
WDI	World Development Indicators
WDR	the World Development Report
WTO	World Trade Organisation

www.ingramcontent.com/pod-product-compliance
Ingram Content Group UK Ltd.
Pitfield, Milton Keynes, MK11 3LW, UK
UKHW021325180426
11947UKWH00017B/1446